Science, Technology and Medicine in Modern History

General Editor: **Professor John V. Pickstone**, Centre for the History of Science, Technology and Medicine, University of Manchester, UK (www.man.ac.uk /CHSTM)

One purpose of historical writing is to illuminate the present. At the start of the third millennium, science, technology and medicine are enormously important, yet their development is little studied.

The reasons for this failure are as obvious as they are regrettable. Education in many countries, not least in Britain, draws deep divisions between the sciences and the humanities. Men and women who have been trained in science have too often been trained away from history, or from any sustained reflection on how societies work. Those educated in historical of social studies have usually learned so little of science that they remain thereafter suspicious, overawed or both.

Such a diagnosis is by no means novel, nor is it particularly original to suggest that good historical studies of science may be peculiarly important for understanding our present. Indeed, this series could be seen as extending research undertaken over the last half century. But much of that work has treated science, technology and medicine separately; this series aims to draw them together, partly because the three activities have become ever more intertwined. This breadth of focus and the stress on the relationships between knowledge and practice are particularly appropriate in a series which will concentrate on modern history and on industrial societies. Furthermore, while much of the existing historical scholarship is on American topics, this series aims to be international, encouraging studies on European material. The intention is to present science, technology and medicine as aspects of modern culture – analysing their economic, social and political knowledge, and how it was shaped within particular economic, social and political structures.

Such analyses should contribute to discussions of present dilemmas and to assessments of policy. 'Science' no longer appears to us as a triumphant agent of Enlightenment, breading the shackles of tradition, enabling command over nature. But neither is it to be seen as merely oppressive and dangerous. Judgement requires information and careful analysis, just as intelligent policy-making requires a community of discourse between men and women trained in technical specialties and those who are not.

This series is intended to supply analysis and to stimulate debate. Opinions will vary between authors; we claim only that the books are based on searching historical study of topics which are important, not least because of the cut across conventional academic boundaries. They should appeal not just to historians, or just to scientists, engineers and doctors, but to all who share the view that science, technology and medicine are far too important to be left out of history.

Titles include:

Lynda Bryder
WOMEN'S BODIES AND MEDICAL SCIENCE
An Inquiry into Cervical Cancer

Alex Mold and Virginia Berridge
VOLUNTARY ACTION AND ILLEGAL DRUGS
Health and Society in Britain Since the 1960s

Biology, Computing, and the History of Molecular Sequencing

From Proteins to DNA, 1945–2000

Miguel García-Sancho
Research Fellow, Spanish National Research Council (CSIC), Madrid

© Miguel García-Sancho 2012

First published 2012 by
PALGRAVE MACMILLAN

Palgrave Macmillan in the UK is an imprint of Macmillan Publishers Limited, registered in England, company number 785998, of Houndmills, Basingstoke, Hampshire RG21 6XS.

Palgrave Macmillan in the US is a division of St Martin's Press LLC, 175 Fifth Avenue, New York, NY 10010.

Palgrave Macmillan is the global academic imprint of the above companies and has companies and representatives throughout the world.

Palgrave® and Macmillan® are registered trademarks in the United States, the United Kingdom, Europe and other countries.

ISBN: 978–0–230–25032–1

This book is printed on paper suitable for recycling and made from fully managed and sustained forest sources. Logging, pulping and manufacturing processes are expected to conform to the environmental regulations of the country of origin.

A catalogue record for this book is available from the British Library.

Library of Congress Cataloging-in-Publication Data

García-Sancho, Miguel, 1978–
   Science, technology, and medicine in modern history :
  from proteins to DNA, 1945–2000 / Miguel García-Sancho
     p. cm.
   Includes bibliographical references.
   ISBN 978–0–230–25032–1 (hardback)
    1. Nucleotide sequence – Data processing. 2. Amino acid
  sequence – Data processing. 3. Genetics – Technique – History. I. Title.
QP620.G37 2012
572'.65—dc23   2012004849

10 9 8 7 6 5 4 3 2 1
21 20 19 18 17 16 15 14 13 12

Printed and bound in the United States of America
by Edwards Brothers Malloy, Inc.

To my family and, especially, my wife Cate
and daughter Clarissa for their infinite patience

Contents

Figures

Acknowledgements

I would particularly like to thank John Pickstone, from the Centre for the History of Science, Technology and Medicine at the University of Manchester (CHSTM), for taking time from his busy schedule to be involved in the editing of this book; Andrew Mendelsohn, my former PhD Supervisor at Imperial College; María Jesús Santesmases, my line manager at the Spanish National Research Council; and Catherine Heeney, my wife and colleague. Their contributions of time, insightful comments and productive working conditions, often at the expense of other pressing professional – and personal – commitments, were invaluable. Palgrave Macmillan and two anonymous referees have provided feedback and continued support. Joanna Baines, also at CHSTM, Manchester, has made adjustments to my English where necessary.

The following people provided valuable help in the production of this book: John Lagnado (UK Biochemical Society Archives); Frances Martin and Joan Green (Sanger Institute Library); Annette Faux and Michael Fuller (Archives of the MRC Laboratory of Molecular Biology of Cambridge); David Edgerton and Max Stadler (Centre for the History of Science, Imperial College London); Soraya de Chadarevian (Center for Society and Genetics, UCLA); Fred Sanger, John Sulston, Sydney Brenner, Bart Barrell, Richard Durbin, and Alan Coulson (Sanger Institute and Laboratory of Molecular Biology of Cambridge); Adam Bostanci, John Dupré, Staffan Müller-Wille, Sabina Leonelli and other researchers at the ESRC Centre for Genomics in Society (University of Exeter, UK); José Manuel Sánchez Ron, Javier Ordóñez, Antonio Sillero and Rafael Garesse (Universidad Autónoma de Madrid); Leroy Hood, Lee Rowen, Tawny Burns, Mike Hunkapiller, Suzanne Beim, Tim Hunkapiller, Lloyd Smith, André Marion, Steve Fung, and Robert Waterston (Institute for Systems Biology, Alloy Ventures, Discovery Bioscience, UW Madison, Applied Biosystems, University of Washington); Shelley Erwin and Bonnie Ludt (Caltech Archives); John O'Neill and Tom Thomas (Applied Biosystems Library and Archives); Caltech and University of Washington Housing Services; Colin Harris (Western Manuscripts, Bodleian Library, Oxford); Olga Kennard (formerly of Cambridge Crystallographic Data Centre); Ken Murray and Alix Fraser (University of Edinburgh); Greg Hamm (GPC-Biotech); Graham Cameron and Mark Green (European Bioinformatics Institute); Bruno Strasser (Yale University); Christina Brandt, Hans-Jörg

Rheinberger and other researchers at the Max Planck Institute for the History of Science; Richard Ashcroft (Queen Mary University); Carsten Timmermann, Mick Worboys, Duncan Wilson and other researchers at CHSTM, Manchester; Edna Suárez (UNAM, Mexico); Ana Romero, Concha Roldán, Javier Moscoso and other researchers at the Institute of Philosophy, Spanish National Research Council.

The Wellcome Trust, the Science Museum, the Biochemical Society, the Nobel Foundation, T. Gossling, L. Smith, Elsevier, Springer, Oxford Journals, SAGE and the Anton Dohrn Zoological Station all kindly granted me permission to reproduce copyrighted material, a process well expedited by the Copyright Clearance Centre. The archives and scientists I interviewed, as listed in the Appendices, allowed me to reproduce unpublished material.

The research on which this book is based was conducted while holding doctoral fellowships awarded by Caja Madrid Foundation, Madrid City Hall and Residencia de Estudiantes (Spain), together with a Hans Rausing Fellowship awarded by the Centre for the History of Science, Technology and Medicine, Imperial College, London. Fieldwork and other expenses were covered by small grants awarded by the Royal Historical Society.

The idea of writing this book arose at the end of a postdoctoral fellowship in History of Medicine awarded by the Wellcome Trust at CHSTM, Manchester. Without that intellectual environment and the enthusiastic support of John Pickstone, it would not have come to print. Preparation of the manuscript was made possible by the generous support of my boss, María Jesús Santesmases, and a postdoctoral JAE-Doc fellowship awarded by the Spanish National Research Council (CSIC), as well as the projects FFI2009-07522 and HUM2006-04939/FISO, awarded to my research group by the Spanish Ministry of Science and Innovation.

Introduction: An Historical Approach to Sequencing

> The history of science, as practiced in the last decades, aims to depict the social and material dimensions of scientific practices in its own context…. The writing of history is supposed to deliver detailed genealogical narratives that show temporal connections between ideas, fields, events, actors, and objects that mediate between past and present. Some of these connections may not be apparent from the viewpoint of the synchronic analysis characteristic of sociological studies. Genomics is even more obscured by the fact that we are dealing with a case in the recent history of science. On some particular issues the temporal dimension does not have enough depth and thus it may be easy to take sociological accounts for historical ones…. Sociological and historical accounts illuminate each other. But in order to be useful to each other, the historical perspective needs to bring into light actors, objects, and spaces that are not revisited from a more contemporary or synchronic perspective.
>
> Suárez-Díaz, 2010, p. 66.

I will never forget the puzzling sensation I experienced during my first visit to the Wellcome Trust Genome Campus. The Campus is located in Hinxton, a small town in Cambridgeshire set within a very traditional British landscape. Having undertaken a long local bus journey, crossing large amounts of the surrounding rural area, the closest stop – and the landmark to avoid missing it – is next to *The Lion,* an old public house with a reputation for fine Sunday roasts. Next to the pub, the entrance to the Campus is marked by a huge black fence. Originally a 55-acre estate surrounding a seventeenth-century mansion, enormous swathes of green land and trees lay before me. As I walked down the Campus's main track, a lake appeared to one side and on the other, a futuristic construction of glass and iron. These buildings house the various laboratories of the Sanger Institute, the first institution to reside on the Genome Campus.

The Sanger Institute, originally the Sanger Centre, was initiated in 1993 as a joint enterprise of the Medical Research Council (MRC) – the body of the UK Government which manages biomedical research – and the Wellcome Trust – a charity, originally deriving from a pharmaceutical company, that, by the early 1990s, had become a major funder of British biomedicine. The Institute entered the peaceful setting of Hinxton when the Wellcome Trust, having purchased the estate, decided to build a Genome Campus, and invited relevant researchers and institutions to inhabit the site. The European Bioinformatics Institute (EBI), a section of the prestigious European Molecular Biology Laboratory (EMBL), also accepted this invitation and relocated to the small British town from Heidelberg. These two institutions, the Sanger Centre and the EBI, then embarked on an audacious enterprise: to contribute to the Human Genome Project (HGP), an ongoing international effort aimed at the determination and computer interpretation of the 3,000 million chemical units which constitute our genetic material. These chemical units are called nucleotides and, according to their chemical composition, are of four types: adenine, cytosine, guanine and thymine.

The HGP was virtually concluded by 2001, with the publication of a draft covering more than 90 per cent of the human genome – the complete set of nucleotides forming our genetic material. The nucleotides are linearly aligned within the DNA molecule, the double-helical structure which had been elucidated some 50 years earlier in nearby Cambridge, thereby providing evidence that DNA is the material of which genes are made. The Sanger Institute and the EBI are currently studying the genome of other species, such as the mouse, the zebra fish and the malaria parasite. This work is partly conducted at the Institute's Sequencing Centre, where dozens of aligned apparatuses determine one-by-one the sequence of adenines, cytosines, guanines or thymines in the genome of the corresponding species. Technicians walk around the machines – called sequencers – ensuring the cycle never stops. At the EBI, computer suites contain dozens of IT 'geeks' visualising the sequential information on screens and building interconnected databases.

This image of man-made efficiency contrasts greatly with the bucolic environment of the Campus. The lake, mansion and fields remind us that things were not always dynamic and automated on the Hinxton estate. Furthermore, the names of the glass edifices – the Morgan and the Sulston Buildings – suggest a complex institutional history. T.H. Morgan was one of the first to investigate the role of genes in heredity. John Sulston became the first director of the Sanger Centre and a major British spokesman for the HGP. Morgan and Sulston's lives ran, respectively, at the beginning and end of the twentieth century, their only connection being a common interest in genes. However, Frederick Sanger, the researcher the Sanger Institute is named after, was not overly interested in genetics till the very end of his career. With his love of gardening and manual laboratory work, he represents the traditional British gentleman scientist, not the modern, cutting-edge whizz-kid of computerised research. His retirement, in 1983, came shortly before automatic sequencing of entire genomes became a scientific and socio-political obsession.

The variety of associations across the Genome Campus is by no means trivial, and they lie at the very heart of this book. The Campus and its location in Hinxton conceal contingencies and aspirations that a historical perspective – often overlooked by its planners and users – will help to uncover. The following six chapters examine when researchers began to sequence biological molecules and how this practice subsequently evolved. They illustrate that the history of sequencing began with proteins rather than DNA and encompasses the entire twentieth century. The history of sequencing thus transcends that of the gene, of sequencing genomes and of building genome campuses.

This long trajectory of sequencing accounts – at times metaphorically – for the contrast between old and new that strikes the visitor to Hinxton. Understanding this contrast, and the anxiety it can produce, allows a critical reassessment of categories which have shaped – and at times limited – the practice of contemporary biomedicine.

A long history writ short

Exploring the history of sequencing is not an original endeavour *per se*. The origins and implications of this practice received much attention in the accounts of the HGP which proliferated following the official launch of the project in 1990, and have flourished again since its conclusion (e.g. Kevles and Hood, 1992; Wills, 1991; Cook-Deegan, 1994; Sloan, 2000). In 2000, the scientific heads of the Project and supporting institutions – among them the National Institutes of Health (NIH) and the US Department of Energy – presented a 'first draft' of the human

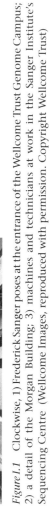

Figure1.1 Clockwise, 1) Frederick Sanger poses at the entrance of the Wellcome Trust Genome Campus; 2) a detail of the Morgan Building; 3) machines and technicians at work in the Sanger Institute's Sequencing Centre (Wellcome Images, reproduced with permission. Copyright Wellcome Trust)

genome sequence. It was a meticulously prepared ceremony covered by the world's media, and included speeches by the US President, Bill Clinton, and Prime Minister of the UK, Tony Blair. The aim was to present the HGP as an entirely successful enterprise, befitting the millions of dollars of investment it had required. Human genome sequencing was also referred to as a fine example of the balance required between competition and cooperation by the two main parties involved: a consortium led by public laboratories from the US, Britain, France, Germany, China and Japan, and a private company called Celera Genomics. President Clinton claimed the sequence would enable an understanding of 'the language in which God created life.'[1] One year later, in 2001, the public consortium and Celera independently published the sequence in the two top scientific journals, *Nature* and *Science* (Bostanci, 2004).

This image of successful, cutting-edge achievement has fostered a growing body of literature on the so-called Ethical, Legal and Social Implications (ELSI) of the HGP a field recently renamed Genomics and Society, so as to include the rapidly diversifying enterprises involved in the accumulation and computer processing of genome data. Scholarship on genomics and society includes investigations from a wide range of perspectives, among them bioethics, politics, economics, law, philosophy and the sociology of biology (for e.g., Brown, Rappert and Webster, 2000; Häyry et al., 2007; Barnes and Dupré, 2008; Harvey and McMeekin, 2007; Parry, 2004; Reardon, 2005). The increasing public interest in genomics and the HGP has also prompted biographies and lay accounts by researchers involved in these enterprises, or the science popularisers who have interviewed them. These accounts include reconstructions of the past shaped by the researchers' experience and are often used to justify their views on a number of scientific and socio-political issues (Cantor and Smith, 1999; Sulston and Ferry, 2002; Venter, 2007; Watson, 2003; Hood, 2003, 2008; Davies, 2002; Brown, 2004; Gee, 2004).

The historical perspective of this literature – which is generally not a major consideration of the authors – tends to follow three main arguments: (1) sequencing is a consequence of the technical revolution that occurred in molecular biology during the 1970s; (2) this revolution was the result of accumulated knowledge surrounding the structure of DNA; and (3) the new, revolutionary techniques of molecular biology paved the way for genomics, bioinformatics, biotechnology and other fresh, promising and interdisciplinary biosciences. Sequencing is, thus, considered to be a technique of molecular biology, a contention that has led to its history being conflated with that of the DNA molecule.

Despite the majority of authors being aware there was a widespread line of research on protein sequencing between the 1950s and 1960s, the protein techniques are presented as, at best, remote antecedents of those employed for DNA (Cook-Deegan, 1994, pp. 59–60; Judson, 1992, pp. 52–3). The authors' determination to *explain* genomics and the HGP leads to a fragmented view of the history of DNA sequencing, and the previous – and implicitly less important – protein techniques.

Historian of biology, Edna Suárez, has shown how this historical framework reproduces earlier, standardised accounts of the development of molecular biology, which emerged as a discipline in the mid-twentieth century, absorbing experimental practices and strategies from preceding fields, including genetics and biochemistry (Suárez-Díaz, 2010). During the 1960s, molecular biology garnered prestige and support from biomedical funding agencies, culminating in the awarding of Nobel Prizes to the self-declared founders of the new discipline. These researchers used their increasing authority to write historical and autobiographical accounts in which molecular biology was portrayed as a discipline mainly engaged in the study of DNA, the *master molecule* from which the secrets of life could be deduced (Cairns, Stent and Watson, 1966; Watson, 1969; Jacob, 1974; Monod, 1972; Crick, 1988).

Historians have demonstrated that this portrayal is an oversimplified view of molecular biology, in which the roles of earlier, contributing disciplines have been marginalised, along with other historically relevant molecules, such as proteins (Abir-Am, 1985, 1991; Morange, 2008).[2] Suárez has argued that, thirty years later, the literature on genomics is facing the same problem, with autobiographies and historical accounts written by reputed genomic researchers tending to portray 'previous developments in the life sciences' as 'inevitably leading to the HGP' (Suárez-Díaz, 2010, p. 68).

Within these accounts sequencing development has a linear and progressive history, initiated by the determination of the double helix of DNA in 1953.[3] James Watson and Francis Crick, at the Cavendish Laboratory in Cambridge (UK), elucidated this structure and used it to challenge the then widely accepted hypothesis that genes were constituted of proteins. Watson, Crick and other researchers devoted the following decade to the decipherment of the so-called coding problem: how DNA, as the genetic material, synthesised – that is, determined the structure of – proteins. The continuity of actors between the double helix and the coding problem, together with Watson and Crick's active role in the promotion of molecular biology, led to the identification of this discipline with progress in DNA research. This projection of molecular biology, as an artificially unified field around DNA, has been the main impetus to the conflation of molecular biology with the history of sequencing.

A major focus of this conflation is the so-called recombinant DNA revolution. During the mid-1970s, the generation of molecular biologists which followed Watson and Crick developed a series of techniques to alter the nucleotide structure of DNA. These tools enabled each strand of the DNA molecule to be cleaved into fragments, and particular fragments to be inserted into the genetic material of another organism – in the initial experiments a DNA fragment from a frog was inserted into a bacterium. DNA sequencing has traditionally been considered part of this recombinant revolution, emerging between 1975 and 1977 and incorporating some of the instruments used in contemporary gene transfer techniques. The fact that DNA sequencing researchers also adapted tools employed in the earlier protein sequencing methods – tools entirely unrelated to recombinant DNA – rarely features in the standard histories.

The identification of sequencing with recombinant DNA facilitates its portrayal as a technique that paved the way to biotechnology, bioinformatics and genomics. Over the late 1970s and 1980s, an increasing number of public laboratories created companies to commercialise products derived from alterations of the DNA molecule. These firms were largely devoted to the creation of new drugs or crops, but they also sought to design sequencing instruments and computer programs to process sequence data (Kenney, 1986). Genomes of increasingly complex organisms were sequenced, a scaling-up of sequencing efforts that led to the initial debates on the feasibility of the HGP in 1985 and 1986 (Sinsheimer, 1989). The proliferation of sequence data from these projects fostered a growing convergence between sequencing techniques and a rapidly developing information-processing instrument – the personal computer (Cook-Deegan, 1994, ch. 18). Such was the association of sequencing with computing that researchers within both the natural sciences, and Science and Technology Studies (STS) argued that biology was becoming an 'information science' (Gilbert, 1992; Hood, 1992; Lenoir, 1999; Zweiger, 2001; Moody, 2004).[4] The factors which fostered the convergence of biology and computing, along with the increasing investment in sequencing projects and technologies, have been – and still are – rather under-researched.

Historians have only recently begun to challenge these narratives. The late arrival of this critical perspective is due to the historiographical approach to genomics having been shaped by the ELSI and STS literature on the HGP (Suárez-Díaz, 2010, pp. 66–7). Sequencing has been regarded as a late 1970s technique and, consequently, too

(a)   **Human Genome Project (1990s and 2000)**

**Biotechnology,
bioinformatics and
genomics (1980s)**

**Recombinant DNA
revolution (1970s)**

**Decipherment of
the genetic code
(up to1967)**

Double helix of DNA (1953)

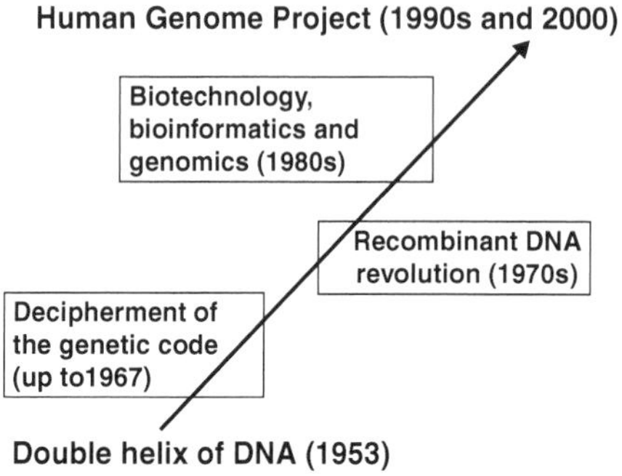

(b)   DNA, RNA and proteins

Any current biology textbook defines DNA as the molecule which constitutes genetic material. Its three-dimensional
structure is a double helix of two coiled threads. Each thread is formed by a row of nucleotides or chemical units of
four types: adenine, cytosine, guanine and thymine. The nucleus of all living cells contains DNA. Its main function
is to synthesiseor determine the structure of proteins, the molecules in charge of most living functions
(muscular contraction, food processing, and so on).

The sequence of nucleotides within DNA directs the organism to synthesisesome proteins or others, according
to the so-called geneticcode. A gene is a fragment of the DNA sequence which determines the structure of one or
various proteins. In order to do so, thedoublehelix of DNA unfolds revealing the relevant gene, which transmits its
sequence to an intermediary molecule, messenger RNA. The messenger RNA carries the information through
the cell'scytoplasm and with the help of another type of molecule – transfer RNA – translates the nucleotide
sequence into a concrete sequence of a mino acids, the chemical units which form proteins.

Sequencing DNA, RNA and proteins is to determine the linear structure of nucleotides or amino acids of these
molecules. The first protein sequencing techniques emerged between 1949 and 1955, and RNA sequencing
through the 1960s. DNA sequencing originated between 1975 and 1977, the decade before the emergence of
biotechnology, bioinformatics and genomics.

From mainframes to personal computers

The first computers originated between the 1930s and 50s, and were characterised by their enormous size.
They were based in sharedcentres of calculation and were alien to both laboratories and homes. Named
mainframes, they were operated via punched cards in which data and processing instructions were coded in a
pattern of perforations through the cardboard surface of the card. Mainframe operators inserted the cards and data
was processed according to the instructions. The results were then sent to the users, sometimes located miles
away from the centres of calculation. Requested operations could be mathematical calculations or classification
of data according to certain criteria.

In the 1960s, minicomputers – cheaper and smaller than mainframes but with a less processing power – emerged.
Mainframes and minicomputers co-existed and in the 1970s, microcomputers and workstations entered the market.
These overcame some of the limitations in data processing of minicomputers and were a suitable size for use in
laboratories and offices by small groups of employees.

Between 1975 and 1981, the first personal computers – Altair, Apple II and IBM's PC –emerged. Their size,
price and operation was adapted to individual use in offices, laboratories and homes. At the same time,
the software industry –initiated in the 1960s –introduced text processors, the programs which currently
constitute the main use of personal computers.

Figures I.2a and I.2b   The 'official' history

recent for a historical approach. A remarkable exception is Soraya de Chadarevian, who has explored the role of protein sequencing in the emergence of the Laboratory of Molecular Biology at Cambridge (UK), founded in 1962 as a pioneer institution in the field.[5] De Chadarevian has suggested a connection between protein and DNA sequencing, implicitly challenging the stories which subordinate this practice to progress in the knowledge of gene structure (de Chadarevian, 1996, 1999). Following her scholarship, a line of research is emerging that proposes new genealogies in the development of sequencing.

A short history writ long

Another pioneering account of sequencing is Michael Fortun's investigation into the development of genomics in the United States. By combining the tools of history and the sociology of science, Fortun has concluded that 'the Human Genome Project does not exist.' He describes a number of dispersed and disconnected initiatives, aimed at detailing the structure of the genomes of different organisms, such as viruses, yeast, the worm *C. elegans* and the *Drosophila* fly, initiatives that preceded the initial discussions about sequencing the human genome in the mid-1980s. The main transformation that human genome sequencing introduced was the 'acceleration' of genomic projects and the creation of an apparently coherent and unified entity – the HGP – which was easier to present to political actors and funding agencies (quotes from Fortun, 1999, pp. 26–7; see also id., 1993, 1998).

This historical perspective has been expanded by Jerôme Pierrel's investigation of the sequencing of RNA, the molecule which mediates in the synthesis of proteins by DNA (see Figure 0.2b). Pierrel has identified an 'RNA age' in the mid to late 1960s, during which this molecule 'seemed to hold more potential than DNA at the laboratory bench.' At that time, RNA sequencing played an important role in attempts to solve the coding problem and was the technique used to tackle the first entire genome of a bacteriophage virus. These viruses infect bacteria and have a simple genome formed by either RNA or DNA. An RNA bacteriophage, MS2, was the first to be fully sequenced, at the University of Ghent in 1976. However, as Pierrel has argued, the parallel emergence of the first DNA techniques led RNA sequencing to fall into oblivion (Pierrel, 2009, pp. 420ff.; see also id., 2012).

The history of protein sequencing and its connections with RNA and DNA techniques is an expanding field. Bruno Strasser has demonstrated that, in the context of biochemical investigations into the relationship between molecular structure and function in the 1950s, the comparison of protein sequences became an important practice. By determining the similarities and differences between amino acids – the chemical constituents of proteins – biochemists hoped to locate the parts of protein sequences involved in functionally relevant processes in the body, such as respiration or nutrient processing. Sequences were also important in initial attempts to solve the coding problem. Computer-assisted comparisons of amino acids from different proteins were made, in the hope that patterns within the DNA sequences in charge of their synthesis could be deduced (Strasser, 2010, 2011).

In the mid-1960s, sequence comparison became a major practice in molecular evolution, a branch of evolutionary biology which aimed to explain taxonomic differences between species through variations in amino acids. Molecular evolutionists systematically compared protein sequences of interrelated species and drew phylogenetic trees in which the processes of differentiation were measured in amino acid changes, fostering the proliferation of databases and computer programs directed at easing sequence comparison. These computer tools were adapted to large mainframe apparatuses, operated with punched cards and located in centres of calculation shared by universities or research institutes (see Figure 0.2b). The historiography of computing has portrayed these mainframes as deriving from mechanical and electrical calculators used in mathematical investigations during the first half of the twentieth century and, especially, World War II (for e.g., Campbell-Kelly and Aspray, 1996, pp. 105ff.). As Strasser has shown, in the post-war years mainframe computers were also applied to the processing of strings of data, such as amino acid sequences (Strasser, 2010, pp. 651–2; see also id., 2006).

Molecular evolution has also been investigated by Suárez, who has demonstrated that the algorithms – or programming commands – used in computer programs to compare protein sequences inspired further software for genome analysis. This has led her to propose a new genealogy between bioinformatics and molecular evolution. Given that molecular evolution did not derive from DNA investigations, this connection acquires strong historiographical relevance, having tended to be overlooked in the autobiographies of molecular biologists or current genomic scientists (Suárez-Díaz, 2009; Suárez-Díaz and Anaya-Muñoz, 2008).

In her research, Suárez seeks to unpack the concept of interdisciplinarity. When the current literature on bioinformatics emphasises the *interdisciplinarity* of this field, it is normally referring to the unacknowledged absorption of prior practices, such as those of evolutionary-oriented protein sequence comparison. Suárez's concern with interdisciplinarity, by contrast, seeks to prompt new narratives: narratives where the motivation is not just a 'sense of justice', but formulating 'problems that demand the attention of a new generation of historians' (Suárez Díaz, 2010, p. 82).

Both Suárez and Strasser have emphasised that computer-assisted accumulation of sequence data preceded the emergence of genomics and bioinformatics. They have described computer programs and mainframe apparatuses far removed from our current perception of informatics and its application to biology. These new genealogies are, however, based on different theoretical and methodological approaches. Whereas Strasser follows the 'historical trajectory' of the 'practices' of sequence collection, comparison and computation, Suárez has focused on the 'mundane factors' of research activity, such as instruments or tools. These approaches have allowed Strasser and Suárez to present a long-term history, one free from the attraction of scientific celebrities or currently glamorous research projects (Strasser, 2010, pp. 651ff.; Suárez-Díaz, 2010, pp. 70ff.).

These emerging genealogies are the starting point for my book. The following six chapters present, for the first time, a complete historical picture of sequencing, one which expands and unifies the research I have discussed. I detail the continuities and changes between protein, RNA and DNA techniques, and explain why the latter monopolised scientific and socio-political favour in the 1980s and 1990s. I also provide a historical rationale for the acceleration of sequencing, and contribute towards attempts to historicise connections between biology and computing. This results in a *longue durée* that complements and enriches those previously created by historians.[6] A major conceptual tool in my research is the comprehension of sequencing as a particular form of work.

Sequencing as a form of work

A common point in the popular, autobiographical and ELSI literature on genomics is the understanding of sequencing as a technique, a method devised to solve a particular experimental problem. Sequencing is portrayed as an instrument which originated and was developed within the experimental pathway of a particular scientific discipline, molecular biology. This understanding leads to the presentation of sequencing as part of the recombinant DNA revolution and to a focus on the generation of molecular biologists which *invented* the new recombinant techniques.

Sequencing can be considered alternatively as the result of the confluence of a variety of practices in a *form of work*. This understanding emphasises the role of both proponents and users of these practices, and places their convergence in a specific historical moment. It is then easier to link sequencing with a multiplicity of actors – molecular biologists or not – who, in finding responses to practical problems, crossed disciplinary boundaries. The category of form of work stands as an intermediate between Strasser's long-standing practices and Suárez's mundane tools. It facilitates a history unrestricted by disciplinary frameworks and highlights both researchers and the instruments they used. Following this line, my book will investigate when biologists began to use sequencing instruments, rather than when sequencing, as a technique, revolutionised molecular biology and paved the way to genomics, bioinformatics and biotechnology.

Sequencing as a form of work connects with the widely influential categories that John Pickstone formulated to analyse the history of science, technology and medicine. From the sixteenth century onwards, according to Pickstone, this history can be explained by the emergence, recurrence and interactions between three ways of 'knowing' and 'working': (1) natural history and craft; (2) analysis and rationalised production, and (3) experimentation and systematic innovation. To these he added the reading of meanings and their use in rhetorics. Pickstone's ways of knowing and working facilitate investigation, not only of the formation of disciplines, but also of features which cross between them (Pickstone, 2000). A history of science, technology and medicine written from this perspective, therefore, differs from one based on the development of traditional academic disciplines.

In a reappraisal of his work, Pickstone proposed 'working knowledge' as a category which unifies his previous ways of knowing and working. He demonstrated how the proliferation of analytical and rationalised working knowledges after the 1970s – including sequencing – has led some researchers to perceive a 'de-realisation of disciplinary separations and the emergence of one biology' (id., 2007, p. 514). Analytical working knowledges reduce entities to data about their elements; their use may require similar skills, independent of the analysed entity. This helps explain why sequencing absorbed practices from a variety of fields and, after the 1960s, interacted with other forms of work directed towards data processing, such as computer programming and database management.

The investigation of recent biomedicine in the light of Pickstone's categories has become a main line of historical inquiry. In a recent collection of essays, Pickstone has renamed *natural history* as *sorting kinds*, in order to emphasise the independence of this way of knowing from disciplinary affiliations, and highlight its crucial role in twentieth century science, technology and medicine (Pickstone, 2011a, p. 236). Strasser and de Chadarevian distinguish between analysis through exemplars and analysis through comparisons, arguing that the history of molecular biology – traditionally considered an essentially experimental discipline – shows how important analytical working knowledges were, namely the study of model organisms and the comparison of protein sequences and structures (Strasser and de Chadarevian, 2011; see also de Chadarevian, 2000). This intermediate category of analysis helps to overcome the rigid separation that both biologists and historians establish between natural history and experimental research. Biologists can analyse entities with the aim of forming data collections, or conduct analytical experiments to determine the elements of an organism and the interactions between them (Pickstone, 2011b, pp. 355ff.). In regard to this, Strasser claims that current biomedicine operates in a 'hybrid culture' in which sorting databanks and conducting experiments are deeply interdependent (Strasser, 2011, p. 61; id., 2010, pp. 652–3). Jean-Paul Gaudillière proposes a new category, 'ways of regulating', to explain the shift from molecular biology to biotechnology and, more generally, the commercial development of drugs (Gaudillière, 2009; Gaudillière and Hess, 2009).[7]

This book will use the history of sequencing to explore why it became a widespread analytical form of work in biomedicine. With the term form of work, I do not intend to propose an alternative category, but to show how Pickstone's working knowledges interact at the bench level in the development of a new way of conducting biomedical research. From the mid-1940s onwards, sequencing evolved from analytical experiments to the systematic analysis of large numbers of specimens and the consequent sorting dilemmas, characteristic of natural history: how to arrange data in such a way that parts can be usefully retrieved and patterns understood. Sequencing also evolved from innovation and craft to rationalised production and large scale mechanisation. With the advent of genomics in the mid to late 1980s, the regulation of sequencing shifted from a marginal position to top priority within biomedicine.

The history of sequencing cannot be fully captured by the standard narratives of biochemistry, molecular biology or biotechnology – or by a rigid distinction between the history of science and technology. It necessitates our putting ways of knowing, working and regulating into historical action.

Structure and methodology

The three parts into which this book is divided address the emergence of sequencing and its subsequent mechanisation, that is, its performance by automatic apparatuses. I adopt a broad definition of sequencing, one which includes both the extraction of sequence data from the molecule – protein, RNA or DNA – and the handling and storage of such data. Since the mechanisation of sequence handling and storage was largely independent from that of sequence determination techniques, I address each process in independent book parts, one devoted to sequencing software and databases, another to automatic sequencers.

Each book part contains two chapters in which sequencing, as a form of work, interacts two-directionally with its historical environment.[8] The historical environment includes people – not only researchers – institutions, disciplines, concepts, values and attitudes towards biomedical research. Academic disciplines, as I discussed, are not useful as analytical frameworks, but do shape the decisions that actors make and are, therefore, historically relevant. I argue that, throughout its history, sequencing has unified apparently different – even opposed – categories, such as technical and scientific work; biology and chemistry; experimentation and computation, or academia and commercialisation. Given that these categories are still largely applied to the practice of biomedicine – and in scientific policy more generally – my history of sequencing calls for their critical reassessment.

In Part I of the book, I address the emergence of sequencing by investigating the career of Frederick Sanger. This British biochemist – after whom the Sanger Institute is named – developed the first methods of determining the sequence of proteins and DNA between 1943 and 1977. By investigating his research trajectory, I will argue that *sequence determination* began as a form of work framed within protein chemistry and, from the late 1950s onwards, gradually incorporated practices from molecular biology. This incorporation was triggered by Sanger's move from the Department of Biochemistry to the Laboratory of Molecular Biology at Cambridge (LMB) in 1962 and the subsequent coinage of the term *sequencing.*

In Part II, I investigate the progressive automation of (1) the assemblage of data from different parts of proteins and DNA into an entire sequence, and (2) the compilation and storage of increasingly large volumes of sequences. I focus on the design of the first sequencing software and databases between the 1960s and 1980s. Sequencing, in this process, incorporated a growing number of practices from professional data management in offices and government departments – practices alien to biology and the academic world at that time. The reconstruction of the history of these practices enables me to demonstrate that the instruments of sequence data gathering, organisation and storage, all originated outside the standard academic careers of biologists. This lack of a clear academic and disciplinary origin qualifies the *interdisciplinarity* that biologists, politicians and science commentators usually attribute to bioinformatics.

In Part III, I explore the development and commercialisation of the first automatic sequencers, especially those designed for DNA sequencing between 1980 and 1998. By analysing the efforts of a group at the California Institute of Technology (Caltech) and the company Applied Biosystems, I show that automation changed the identity of sequencing as a form of work. The manual skills and dexterity required in Sanger's techniques were substituted by large-scale sequencing projects, founded on rational organisation and efficiency. I argue that this new identity should not be seen as a natural or inevitable step in the history of sequencing. Between the late 1980s and early

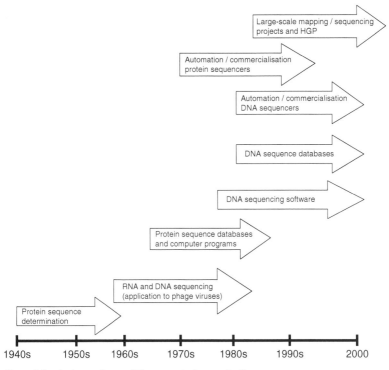

Figure I.3   A chronology of the reported case studies

1990s, the researchers involved in manual sequencing challenged full automation, proposing an approach based on permanent human control over the sequence data.

In all the case studies presented, I have started with oral histories and a review of the autobiographical accounts offered by the researchers involved. These materials have been approached as 'historically meaningful' narratives in which researchers have reconstructed the past according to their current interests and views (Abir-Am, 1991, Abir-Am and Elliot, 1999; Suárez-Díaz, 2010). The actors' accounts have subsequently been submitted to critical historical analysis, using scientific publications, secondary literature and unpublished sources. The latter include correspondence, memos, laboratory notebooks, grey literature – periodical internal reports of research institutions – corporate archives, patents and funding applications, many of which are still uncatalogued and in the hands of the researchers involved. Following an approach developed by F.L. Holmes (2001, 2006), I offer an alternative account of the events in light of the records investigated. I also reflect on the implications of the actors' narratives for the expectations and current socio-political view of the progress of biomedicine. A full list of archival sources and oral histories can be found in the Appendices.

The fact that many of my target actors and institutions are still active has enabled me to enter into these environments and interact with working researchers. Periodical visits to the EBI, LMB, Caltech and Applied Biosystems allowed me to immerse myself in their differing research cultures. I also learned the basics of DNA sequencing, both through observation and a short technician apprenticeship in a small laboratory. However, although this brings my research close to ethnographic approaches in the field of genomics and society, my methodology is mainly historical and I have not directly employed sociological tools. A more specific collaboration with ethnographers may be a useful avenue for future investigations. Sociologists of contemporary biomedicine are increasingly introducing a historical perspective into their research and, conversely, there have been attempts to construct a sociologically-informed history (Brown, Kraft and Martin, 2006; Keating and Cambrosio, 2006; Parry, 2004; Abir-Am, 1987; Harwood, 1993). This scholarship still includes little cooperation between sociologists and historians. As in biomedical research, institutional, personal and theoretical differences need to be overcome before a real – rather than rhetorical – convergence can be achieved.

Part I

Emergence: Frederick Sanger's Pioneering Techniques (1943–1977)

Of the three main activities involved in scientific research, thinking, talking, and doing, I much prefer the last and am probably best at it. I am all right at the thinking, but not much good at the talking. Doing for a scientist implies doing experiments, and I managed to work in the laboratory as my main occupation from 1940...until I retired in 1983. Unlike most of my scientific colleagues, I was not academically brilliant. I never won scholarships and would probably not have been able to attend Cambridge University if my parents had not been fairly rich; however, when it came to research where experiments were of paramount importance and fairly narrow specialization was helpful, I managed to hold my own even with the most academically outstanding.

Sanger, 1988, pp. 1–2.

It is an interesting paradox that so leading a scientific figure as Frederick Sanger has received little attention in popular and academic literature on science, technology and medicine. Sanger, whose pioneering techniques enabled the determination of the sequences of proteins, RNA, and DNA, is one of the very few scientists to be awarded the Nobel Prize twice, in 1958 and 1980. Despite this, only two historians of science have studied his career in any depth (de Chadarevian, 1996, 1999; García-Sancho, 2010).

There are a number of small summaries of Sanger's scientific trajectory in popular literature, mainly in works concerned with the Human Genome Project (HGP). These accounts portray Sanger as a *precursor* of genomics, as his techniques enabled the subsequent determination of a growing number of DNA sequences. His previous work on proteins is represented merely as a remote antecedent, and that on RNA barely mentioned (e.g. Judson, 1992, pp. 51–3; Wills, 1991, pp. 29–33; Cook-Deegan, 1994, pp. 59–61). Sanger's contributions are usually framed in the broader history of progress towards the resolution of the structure of the gene, from the double helix of DNA to a more detailed nucleotide sequence (see Introduction). However, others' achievements in the exploration of DNA's structure have tended to receive far more praise and attention. James Watson and Francis Crick, discoverers of the double helix in 1953, have been the subjects of biographies and autobiographies ever since (Olby, 2009; Watson, 2010).

Sanger has never written a book-length autobiography, or been the subject of a biography. The most extensive account of his career is his paper 'Sequences, sequences and sequences', which illustrates his modesty and technical rigour. In this piece, Sanger insisted that his methods were mainly dependent on the availability, at each historical moment, of instruments for treating protein, RNA or DNA molecules. The various techniques for cutting molecules, separating the resulting fragments and analysing them were, according to him, essential for the three key achievements of his career: (1) the determination of the sequence of insulin (1955), (2) of various RNAs (1960s) and (3) the invention of the first techniques to sequence DNA (1975–77). Interviews given by Sanger, and accounts by his collaborators, have spread this idea of DNA sequencing as largely the result of technological progress (Lagnado, 2005; Hartley, 2004; Stretton, 2002). That the British reference centre for genomics is named the Sanger Institute seems to strengthen this argument and the perception of continuity between Sanger's work and large-scale sequencing projects.

These accounts have been partially qualified by Soraya de Chadarevian, who in two pioneering historical papers stressed the importance of the research environments that surrounded the sequencing techniques. Sanger's professional contacts and institutional settings were essential for the development of sequencing, especially in 1962, when he made a crucial career shift. Leaving the Department of Biochemistry at the University of Cambridge – where he had been working since the early 1940s – Sanger moved to the recently founded Laboratory of Molecular Biology, based in the same city. This move coincided with Sanger's decision to apply his techniques to RNA and DNA, leaving proteins to other members of his group. De Chadarevian argues that contacts with molecular biologists were decisive in Sanger's shift, and equally, that the introduction of sequencing into a molecular biology laboratory shaped the emergence of this discipline (de Chadarevian, 1996, 1999).

De Chadarevian's papers thoroughly analyse Sanger's first protein sequencing methods and their incorporation into the technical repertoire of molecular biology. There is, however, little attention to Sanger's further work on RNA and DNA sequencing, as de Chadarevian's main motivation for analysing Sanger was to investigate the emergence of molecular biology in Cambridge and the role of previous disciplines and techniques – biochemistry and protein sequencing – in that process (id., 1996, especially pp. 382 ff.).[1] In her latest paper, de Chadarevian argues that Sanger's early protein techniques 'informed initial attempts at nucleic acid sequencing', but says little about how the practices and working strategies behind sequencing evolved from proteins to RNA and DNA (de Chadarevian, 1999, quote from p. 203).

In Part I of this book, I address Sanger's entire career (1940–82) with the aim of investigating the emergence of sequencing as a form of work which travelled across institutional and disciplinary boundaries. This investigation requires revisiting retrospective accounts in light of Sanger's scientific papers and, crucially, his laboratory notebooks, which were donated to the British Biochemical Society and are currently available at the Wellcome Trust Archives in London. The next two chapters build on previous work in which I presented sequencing as a practice with a history beyond those of biochemistry, molecular biology and genomics (García-Sancho, 2010). This reappraisal not only highlights the historical specificity of sequencing, but also complements existing literature on the 'molecularisation' of the life sciences (Kay, 1993; Abir-Am, 1982; Rheinberger, 1997; de Chadarevian and Kamminga, 1998). The emergence of sequencing was the result of a growing focus on molecular structure in biology during the mid-twentieth century, and its early history sheds new light on the transformation of biochemistry and the rise of molecular biology between the 1940s and 1970s.

1

The Sequence of Insulin and the Configuration of a New Biochemical Form of Work (1943–1962)

The Department of Biochemistry at Cambridge, where Sanger began his career, was the creation of its first director, Frederick G. Hopkins. Hopkins had arrived at Cambridge when medical sciences were dominated by the physiologist Michael Foster and, during the first decades of the twentieth century, had struggled to establish biochemistry as a new discipline at the University. Hopkins was awarded the first Chair in Biochemistry in 1914, and in 1923 given his own building, the Dunn Institute of Biochemistry. His research agenda, as in physiology, was concerned with the functional processes of the body. Hopkins, however, focused on their molecular basis through a 'dynamic' study of metabolism (Geison, 1978; Weatherall and Kamminga, 1992).

In the subsequent decades, the Dunn Institute came to represent a general and biological approach to biochemistry, in contrast to more medical orientations found elsewhere (Kohler, 1982, pp. 73–89; Holmes and Heilbron, 1992). The Institute's lines of research covered a wide array of aspects of metabolism: the chemical reactions by which a cell processes nutrients and other inputs. Researchers as diverse as Ernest Baldwin, J.B.S. Haldane and Joseph Needham developed their varied careers at the Dunn Institute, focusing respectively on comparison between the metabolisms of different species, genetic investigations of the regulation of the metabolic process, and the development of this process from embryo to adult (Paul, 1984; Abir-Am, 1982, pp. 361–7; Armon, 2012; Strasser, 2010, pp. 626ff.).

Sanger moved to Cambridge from Gloucestershire in the late 1930s in order to pursue an undergraduate degree. He was the second son of a relatively wealthy family – his father was a physician – but was orphaned early in life. His undergraduate study was supervised by Baldwin and coincided with the golden years of metabolic biochemistry. Inspired by his Quaker family background, Sanger refused to perform military service during World War II and was recognised as a conscientious objector (Sanger, 1988; Lagnado, 2005).

In 1940, Sanger began a PhD at the Dunn Institute on the metabolism of lysine, one of the twenty amino acids that constitute proteins. Proteins were by then a privileged object of study, especially the *enzymes*, which catalyse the chemical reactions of the cell. Most biological researchers at that time – including Haldane – believed that enzymes constituted the genetic material which directed the main processes of the body. A number of changes occurred at the Institute during the course of Sanger's doctoral research, which helped foster a new perspective on the study of proteins.

1.1   Chibnall, Fischer and the chemical analysis of proteins

When Sanger arrived at the Dunn Institute, Hopkins was advanced in years, many of his collaborators were leaving and his influence was decreasing; he retired in 1943. Hopkins was replaced by Charles Chibnall, an emerging authority on plant biochemistry from Imperial College, London, whose research focused on amino acid analysis, a chemical technique for investigating the composition of proteins. Chibnall had moved to Cambridge with most of his London team and concentrated on the amino acids of the protein insulin. Upon his arrival, the Dunn Institute began to be known as the Department of Biochemistry of Cambridge (Synge and Williams, 1990; Fruton, 1992, pp. 35–6; Gay, 2007, pp. 148–9 and 171–2).

Historian of biochemistry, Robert Kohler, has shown how Chibnall's appointment substituted Hopkins's general and biological approach with 'much narrower interests' in the chemical investigation of proteins (Kohler, 1982, pp. 88–9). The dynamic and multi-perspective study of metabolic processes was gradually replaced by amino acid analysis as the Department's privileged line of research. Given the importance of insulin for the treatment of diabetes, Chibnall's work was partially supported by a number of British pharmaceutical companies, including ICI and Eli Lilly (de Chadarevian, 1996). The main supporter of biochemistry in Cambridge, however, was the Government's Medical Research Council (MRC).

Chibnall described the practicalities of his research in a 1942 lecture, written shortly before his move to Cambridge. Proteins were broken into their constituent amino acids through different procedures, and chemical reagents were applied in order to determine the quantity of each amino acid. Using this method, he had been able to determine the number of 'aspartic acid, glutamic acid, arginine, histidine and lysine' in a number of proteins. These studies, Chibnall claimed, would allow the correlation of the 'composition' of proteins with the way they acted in the body (Chibnall, 1942, p. 137).

Chibnall placed his work within a research tradition that had arisen in the mid-nineteenth century – mainly in Germany – whereby the methods of organic chemistry were applied to the study of biological molecules. In the same 1942 lecture, he named Justus von Liebig, Franz Hofmeister and Emil Fischer as major influences on his work (ibid., pp. 136–7). Historians of biochemistry have considered these researchers to be among the pioneers who extended the analytical techniques of chemistry to biology (Florkin, 1972, vol. 30, chs 7 and 15; Fruton, 1972, ch. 2; Kohler, 1982, ch. 2). Additionally, Hofmeister and Fischer are credited as postulators of early hypotheses concerning protein structure.

In 1902, based on their chemical analysis of a large number of proteins, Hofmeister and Fischer independently concluded there were recurrent patterns in protein structure. Whereas Hofmeister claimed that proteins were characterised by repetitive chemical links, Fischer argued they were 'chains' of chemically linked amino acids. Fischer further established the 'peptide bond' as the recurrent and characteristic chemical link of proteins, and referred to peptides as the basic protein chains. Proteins were, therefore, molecules formed by many associated peptide chains (Fruton, 1985, pp. 315–16, 326–30; Hofmeister, 1902; Fischer, 1902).

Up to the late 1930s, there were reasons to question these hypotheses. Fischer himself had expressed doubts about the validity of his peptide chains in longer proteins, and studies in other research fields were suggesting alternative structures. One of these was X-ray crystallography, a discipline derived from physics, which at that time was beginning to focus on the three-dimensional conformation of proteins and other biological molecules. During the first decades of the twentieth century, it was generally considered that to explain 'the specific biological and physical properties' of proteins the peptide bond was insufficient. Other types of chemical bond were necessary, giving the molecule a globular structure (Fruton, 1979, quote p. 10). There was also an enduring influence derived from the mid-nineteenth-century field of colloidal chemistry, whose proponents claimed proteins lacked a specific molecular structure, being instead complex mixtures of materials suspended in fluids – such as blood – which did not obey the laws of solution chemistry (Florkin, 1972, vol. 30, pp. 279–85; Creager, 1998, pp. 111–16).

During the 1940s, the debate was being gradually resolved in favour of the molecular structure of proteins. The work of Chibnall and other protein chemists, together with findings in the emergent field of physical chemistry (Edsall, 1979), had demonstrated that, even in long protein chains, a linear arrangement of discrete amino acids was the most plausible structure. Nevertheless, even amongst the supporters of Fischer and Hofmeister, there were disagreements on the composition of amino acids within the peptide chains.

1.2   Periodical or undetermined chains?

Fischer's peptide hypothesis was interpreted in a number of ways during the 1930s and 1940s. One widespread view was that amino acids in the protein chains were repeated at intervals. In 1934, British crystallographer William Astbury, having studied the structure of wool and other animal fibres at the University of Leeds, stated that it was 'uncommon (…) for molecules to be disorganised in the solid state.' Proteins were thus likely to be 'not simply very long, but (…) also periodic polypeptide chain systems' (Astbury, 1934, pp. 15 and 23).

Another proponent of periodicity was Max Bergmann, a former collaborator of Fischer and director of the Kaiser Wilhelm Institute for Leather Research in Dresden. Being Jewish, Bergmann had to flee the Nazi regime in Germany, and made his way to the Rockefeller Institute, New York, where he was appointed Head of the Laboratory of Chemistry in 1934. He created a school devoted to the chemical analysis and synthesis of proteins (Clarke, 1944, pp. 169–70).

Following a series of observations by his assistant Carl Niemann, Bergmann proposed the formula $2^n \times 3^m$, which he alleged described the total number of amino acids in any given protein. The powers (n and m) of 2 and 3 defined the different intervals at which the amino acids were, supposedly, repeated. Bergman and Niemann proposed 'a classification of the numerous individual proteins' in which their basic properties and structure would be mathematically estimated through variations of the formula $2^n \times 3^m$ (Bergmann and Niemann, 1938, quote p. 582; Fruton, 1979, pp. 13 ff.).[1]

Not all researchers, however, interpreted Fischer's theory in the same way. Among biochemists the most critical of periodicity were those based in Cambridge, who declared that proteins were undetermined structures and their study required chemical analysis, not mathematical predictions. Norman W. Pirie and Albert Neuberger, Sanger's PhD supervisors, were among the first to attack Bergmann and Niemann during the late 1930s, employing mathematical arguments to do so. In independent papers, they demonstrated that almost every possible distribution and total number of amino acids in a protein might be expressed as a result of the formula $2^n \times 3^m$ (Neuberger, 1939, pp. 25–6; Pirie, 1939, pp. 351–3).

Chibnall also denounced the periodicity hypothesis, using chemical analysis rather than mathematical methods. In his 1942 paper, he demonstrated that the amino acid compositions of edestin, β-lactaglobulin, egg albumin, and insulin, all yielded values that contradicted Bergmann and Niemann's formula (Chibnall, 1942, pp. 139–57). Chibnall concluded that the protein molecule was a 'system of peptide chains of varied composition' (ibid., p. 158) and that, to determine its structure, it was necessary to analyse it physically and chemically, rather than mathematically:

> Clearly the analyst (...) can contribute but little on his own, and there is need for more co-operation with the physical chemist and crystallographers, who are able to investigate the properties and structure of (...) the intact protein molecule itself. (...) Meanwhile, I think that those interested in proteins would be wise to regard the Bergmann–Niemann hypothesis as still tentative and in any case as applicable only to the component peptide chains of the molecule, for much of the evidence hitherto brought forward to support it has been based on inadequate experimental data and has demonstrated nothing more than the hypnotic power of numerology. (ibid., p. 159)[2]

This context, with the critique of periodicity and a shift from a metabolic to a chemical approach to biochemistry, framed Sanger's early work on insulin during the mid and late 1940s. Taking Chibnall's amino acid analysis, he substantially modified the technique, allowing it not only to determine the composition of proteins, but also the 'sequence' of their amino acids along the chain (e.g. Sanger, 1949a,b).

1.3   Determining an unpredictable 'sequence'

Chibnall's arrival in Cambridge coincided with the completion of Sanger's PhD thesis. Their similar personalities, marked by discretion, an interest in techniques and a preference for the bench over the meeting room, led to a mutual affinity (Chibnall, quoted in Fruton, 1992, pp. 35–6). Chibnall invited Sanger to participate in his project on the chemical analysis of insulin – research that, having been initiated at Imperial College, had become the main focus of his group. Accepting the offer, Sanger was incorporated into the team. Instead of merely applying the established amino acid analysis techniques, however, he introduced a series of variants that determined not only the number and nature of the amino acids in the protein, but also their position.

Sanger's first assigned duty within the insulin project was to quantitatively estimate and identify those amino acids at the end of the molecule's chains (Sanger, 1988, pp. 4–5). To do so, he devised a method that combined established techniques with new instruments. Among the established techniques was column chromatography, a method that enabled separation of those amino acids cleaved from the protein chains (Morris, 1998; Gordon, 1977). This method had been introduced in the early 1940s and extensively applied to amino acid analysis by Chibnall's team, as well as other researchers (Sanger, 1988, p. 5; Martin, 1952, p. 366).

Sanger, however, added to the method a chemical substance – dinitrofluorobenzene (DNFB) – advocated by Chibnall but not previously utilised in protein analysis (Chibnall, quoted in Fruton, 1992, p. 36; de Chadarevian, 1999, p. 203). DNFB reacted with the last amino acids of the insulin chains, dyeing them yellow. This substance allowed Sanger, in line with Chibnall's request, not only to quantify, but also to identify the amino acids. By applying DNFB and cleaving the insulin molecule with acids, he broke all the amino acids of the chains. He then separated the mixture of loose amino acids through column chromatography and determined, from their yellow colour, which ones were located at the end of the chains (Sanger, 1945, pp. 507–15; 1988, pp. 4–6).

Sanger reported the results of those experiments in 1945. In the paper, he defined insulin as a molecule formed by 'four open polypeptide chains' (id., 1945, p. 514).[3] He was, therefore, describing the protein as a series of peptides linked in amino acid chains, in line with Chibnall's investigations and Fischer's hypothesis. In his subsequent two papers, both published in 1949, Sanger began referring to the insulin molecule as a 'sequence', a term which, despite having been used before, had never appeared in Fischer or Chibnall's articles (id., 1949a, p. 563). Sanger's sequences were formed by the ends of the insulin chains together with the neighbouring amino acids.

Between 1945 and 1949, Sanger identified and determined the position of the four to five amino acids beside those at the ends of the insulin chains. He achieved this by introducing two innovations into his technique. Firstly, the protein was submitted to chain cleavage – known as hydrolysis – partially rather than completely. Instead of separating the amino acids individually, Sanger reduced the time of acid exposure and obtained peptides, mini-chains of four to five amino acids (ibid., pp. 564–5; id. 1949b, pp. 154–7). Secondly, Sanger incorporated an improved separation method called paper chromatography. In this method the medium was a sheet of cellulose, which enabled the separation to be performed in two directions instead of one. The resulting amino acid pattern on the paper – the chromatogram – was, therefore, two-dimensional (id., 1949a, pp. 564–5; 1949b, pp. 154–7; see Figure 1.3).

This new method, which Sanger named the *degradation approach,* consisted of submitting various insulin samples to partial hydrolysis and the resulting yellow peptides – DNFB labelled and at the end of the chains – to further cleavage. In this way, he obtained a series of overlapping amino acid fragments, for instance phenylalanine, phenylalanine-valine and phenylalanine-valine-aspartic acid. By separating the fragments through chromatography and submitting them to amino acid analysis – as Chibnall and his group members were doing – it was possible not only to identify and quantify, but also to deduce the order of amino acids. 'The increasing complexity of each of those peptides', Sanger claimed, 'suggested that they were all breakdown products of the same peptide chain' (Sanger, 1949a, p. 556; 1949b, pp. 154–7, quote in 155; Wills, 1991, p. 30).

Sanger referred to these amino acid peptides as *sequences*. His use of the term here differed from that of the term *chain* by Fischer, Chibnall, Sanger himself (in his 1945 paper), and other members of his group. Sanger, nevertheless, used his sequences as evidence for the protein structure hypothesis preferred in Cambridge. In the conclusions of one of his 1949 papers, he stated that the insulin investigations, and comparison of the resulting terminal peptides with those of other proteins, suggested there were no rules governing protein structure and it was, consequently, unpredictable:

Investigations of the free amino groups of a number of proteins by [this] technique (...) have shown that the terminal position in the protein chains may be occupied by a variety of different amino acids. There appears to be no principle that defines the nature of

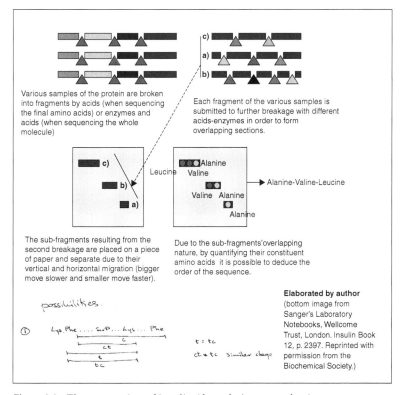

Figure 1.1   The sequencing of insulin (degradation procedure)

the residue occupying this position in different proteins and it would seem probable that this would apply to other positions in the molecule. (Sanger, 1949a, p. 573)

Sanger's early experiments on insulin were therefore framed by the perspectives and research interests of Chibnall's group. They were, however, progressively shifting towards new goals. On the one hand, Sanger exclusively used chemical methods to analyse protein chains, and suggested these were unpredictable, in line with the views of the Cambridge Department of Biochemistry.[4] On the other, in so doing he gradually abandoned his first duty of determining the terminal amino acids of insulin and directed his methods to analysing increasingly larger *sequences* within the molecule. Sanger's latter goals resulted in a new and distinctive form of work within biochemistry, identified during the 1950s as sequence determination.

1.4   An emerging form of work

By introducing further method modifications, Sanger expanded his insulin sequences. He began combining acids with enzymes in the cleavage of the protein chains, to break the molecule at specific sites – such as between valine and alanine – revealing the first amino acid at the beginning and end of each fragment. Thus, the degradation procedure could be applied not only to the edges, but to the central parts of insulin (Sanger and Tuppy 1951a,b; Sanger and Thompson, 1953a,b; Sanger, 1959; 1988, pp. 9–11).

Between 1951 and 1955, Sanger was able to determine the entire sequence of insulin using this method. During this period, Sanger increasingly employed the term *sequence* in his papers. Whereas in previous articles this term had been either absent or combined with other ways of referring to protein chains, such as 'free amino groups', 'terminal peptides' or insulin 'structure' (Sanger, 1945, 1949a,b), the titles of these new papers clearly stated that the object of Sanger's work was the 'amino acid sequence' of the molecule (Sanger and Tuppy, 1951a,b; Sanger and Thompson, 1953a,b). Furthermore, Sanger declared his intention 'to determine the complete amino acid sequence' of the different chains of insulin, which by that time had been reduced from four to two (quote in id., 1953b, p. 23).

Sanger was assisted by two biochemists in the determination of the insulin chains: Hans Tuppy helped him with the phenylalanyl chain between 1949 and 1951, and E.O. Thompson worked on the glycyl chain between 1951 and 1953. They co-authored papers with Sanger, learned the sequence determination techniques and disseminated them in a variety of ways during their careers.[5] The last two years of the insulin effort were devoted to determining the disulphide bridges that united the chains.

Sanger was not the first to use the term sequence when referring to protein structure. In 1943, Archer Martin and Richard Synge, biochemists at the Wool Industries Research Institution in Leeds, had raised the possibility of studying 'structural sequences' in proteins (Gordon, Martin and Synge, 1943). At that time, protein chemistry, like X-ray crystallography, seemed a promising tool for the quality control and improvement of commercial natural products such as wool, leather

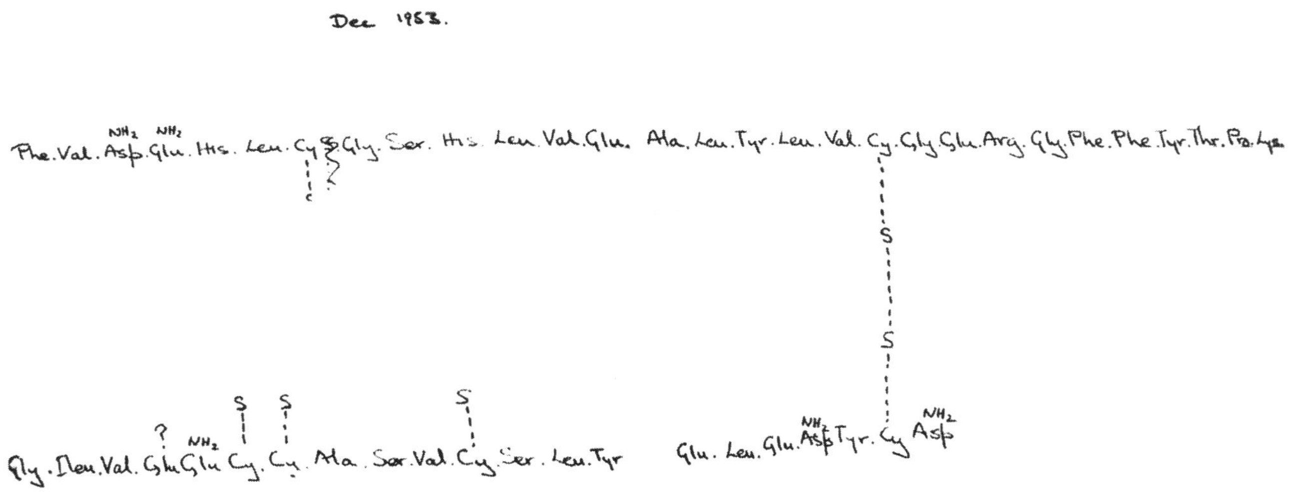

Figure 1.2 Sanger's advanced draft of the sequence of insulin, dated December 1953. The still hypothetical disulphide bridges are represented by the letter 'S'. (Sanger's Laboratory Notebooks, Wellcome Trust, London, Insulin Book 9, unnumbered pages. Reprinted with permission from the Biochemical Society.)

or insulin. It was not unusual for researchers in those fields, such as Astbury, Chibnall and Bergmann, to co-operate with industry or work in industrial laboratories.

Despite Sanger not being industrially minded, his career significantly intersected with those of Martin and Synge. Synge had also been supervised by Pirie at the Dunn Institute of Biochemistry during the second half of the 1930s, shortly before Sanger's arrival. Martin had begun his career at an offshoot of that institute, the Dunn Nutritional Laboratory, created to address problems resulting from the processing of nutrients. Such problems were a research priority in the inter-war years, especially after Hopkins's contribution to the discovery of vitamins as metabolic factors (Weatherall and Kamminga, 1992, pp. 64–6 and 83; Kamminga and Cunningham, 1995, chs7–9). Synge and Martin, like Sanger, were focused on the development of techniques.

In Leeds, Martin and Synge's work addressed column and paper chromatography, the main techniques for the experiments Sanger reported in his 1945 and 1949 papers (Martin and Synge, 1941; Consden, Gordon and Martin, 1944; Sanger, 1945, 1949a, 1988). Sanger and Synge met at Arne Tiselius's laboratory in Uppsala during a joint visit in 1947. This laboratory was a major centre for separation techniques and periodically welcomed guest researchers from the late 1930s onwards (Gordon, 1977; Smith, 1977; Pedersen, 1983; Kay, 1988). During his stay, Sanger became familiar with ionophoresis, another separation method based on the electrical charge of amino acids.[6]

The investigations of Martin and Synge, like those of Sanger, had similarities with the work of Chibnall's group. When they first proposed addressing the structural sequences of proteins, their separation techniques were directed at determining only the 'composition' of the small peptide gramicidin: the chemical identification and quantitative estimation of its amino acids. The determination of the sequence was only stated as a long-term possibility (Gordon, Martin and Synge, 1943). However, with the introduction of paper chromatography in 1944, Martin and Synge presented this new separation technique as also enabling the 'qualitative analysis of proteins' (Consden, Gordon and Martin, 1944, p. 224). Three years later, they used paper chromatography to determine the 'sequence' of gramicidin and proposed applying the technique 'with increased confidence for studies of the sequence of amino acid residues in peptide structures generally' (Consden et al., 1947, p. 602).

Sanger had also been working on gramicidin during the 1940s, alongside his early insulin studies. Prior to 1949 – when he first used the term *sequence* – Sanger had begun defining the results of paper chromatography as 'qualitative'.[7] These parallelisms and the crucial use of paper chromatography in the insulin experiments suggest that Sanger adopted the concept of sequence from Martin and Synge. When in the 1950s he began referring to the determination of the complete sequence of insulin, Sanger extensively cited these two researchers in his papers (Sanger, 1951a, p. 463; 1953a, p. 353).

The use of the term sequence and the practices thus embodied in the investigations by Martin, Synge and Sanger, demonstrate that a new form of work was emerging during the 1940s, and consolidated over the next decade with the completion of insulin. Martin and Synge introduced the term sequence, and the practice of determining it, but only as a means to test their separation techniques. Sanger, by contrast, made sequence determination the focus of his career: his research objects were sequences and the methods of determining them included other biochemical tools, as well as separation techniques.

One sees here the emergence of a form of work which was to become characteristic of molecular biology and of interest to almost all branches of biomedicine (Pickstone, 2007, pp. 513 ff.). At the time of its emergence, however, it was just a form of protein analysis confined to a specialised team of protein chemists. Sanger's sequence determination remained strongly framed in the chemical research tradition represented by Fischer and embodied in Chibnall's group during the first half of the 1950s. His determination of insulin was, on the whole, accomplished through the practice of organic chemistry.

1.5   Absorbing (bio)chemical practices

Despite their differences, Bergmann and Chibnall, together with Sanger, and Martin and Synge, shared a working tradition marked by the application of the instruments of chemistry to biological molecules. This tradition had been established by, among others, Fischer – a chemist by background – who, between the late nineteenth and early twentieth centuries became increasingly interested in proteins (Fruton, 1985). The application of chemistry techniques to biology was one of the routes by which biochemistry – or more specifically protein chemistry – acquired its identity as a discipline (Kohler, 1982). It was this identity – a stark contrast to the multi-perspective, metabolic studies of Hopkins's – that framed the initial attempts to determine the sequence of proteins.

In a 1907 lecture, Fischer reflected on this convergence of chemistry and biology, a convergence he considered natural, citing other nineteenth-century German chemists as the precursors of this union. Germany had been – and still was at that time – a leading scientific and industrial power, not only in chemistry, but also in other disciplines. Fischer believed chemistry was equipped with a 'powerful armoury of analytical and synthetical weapons' which would bring a 'clearer insight' into the 'processes which constitute animal and vegetable life' (Fischer, 1907, p. 1765). These methods, developed over the late-eighteenth and nineteenth centuries, were characterised by, respectively, breaking molecules into fragments – analytical – and assembling such fragments into new compounds – synthetic.[8]

Fischer, having analysed and synthesised a large number of compounds, had pioneered the application of these techniques to proteins (Fruton, 1985, pp. 326–30). He named his analytical techniques 'degradation methods' and, as Sanger and Chibnall would do, focused them on the 'composition' of molecules (Fischer, 1907, pp. 1761 and 1753). In his 1907 lecture, Fischer described his experiments as follows:

> The conclusions which have been drawn in other cases from the results obtained by the dissection of compounds have been too frequently confirmed by their synthesis. It is now possible to make this claim on behalf of the proteins, as it has been found to be possible, by a process the reverse of hydrolysis to associate amino acids in such a manner that substances are produced which (...) resemble proteins. I have termed these synthetic products polypeptides (ibid., p. 1761).

Bergmann and Chibnall followed in Fischer's analytical and synthetic tradition, the latter with the research group that relocated to Cambridge in 1943 (Chibnall, 1942, p. 137). Bergmann – formerly a close associate of Fischer[9] – created a school at the Rockefeller Institute where, from the 1930s onwards, a number of young researchers were trained in the synthesis and structural analysis of proteins (Clarke, 1944).

In his history of biochemistry, Kohler has identified 'three distinct styles' characterising the diversity of this discipline during the first decades of the twentieth century: (1) clinical (2) biological and (3) bioorganic-biophysical (Kohler, 1982, p. 7). The work of Fischer, Bergmann and Chibnall, with its strong analytical and synthetic approach, was firmly embedded in the third style. None of these researchers displayed any concern with medical matters or an interest in the biological processes of living organisms, such as metabolism.

Chibnall's move to Cambridge may, therefore, be interpreted as a shift at the Dunn Institute of Biochemistry from a biological to a bioorganic and biophysical style. Lacking the broad and dynamic approach to metabolism of Hopkins, Chibnall favoured the application of the analytical techniques of chemistry to proteins.[10] His model was short-lived, however, since in 1949 Chibnall resigned, weary of his administrative duties and the Institute – by then universally named the Department of Biochemistry of Cambridge – took a more medical orientation (Kohler, 1982, p. 89; Synge and Williams, 1990 p. 81; Weatherall and Kamminga, 1992, pp. 83–4).

Sanger's attempts at sequence determination were decisively shaped by the growth of protein chemistry in Cambridge, and remained an important line of research following Chibnall's resignation. The degradation procedure was based on the breakage of insulin into fragments and its reconstruction through the identification of overlaps. In so doing, Sanger was applying the cleaving, separation and quantification methods of analytical chemistry already used by the other members of Chibnall's group. He was also employing, though not experimentally, the synthetic chemistry strategy of attempting to reconstruct the original molecule from the fragments' overlaps.

Sequence determination, as a distinctive form of work, also added specific features to the protein chemistry of Chibnall's group. By incorporating new instruments into amino acid analysis – DNFB and paper chromatography – Sanger gave a new qualitative dimension to this method, which was then able to yield not only the *composition* of insulin, but also the *sequence*. He also created, with Martin and Synge, the concept of sequence and the aim of achieving it through the development of techniques. Sanger's sequence determination survived both Chibnall's leaving and the shift in Cambridge to a medically-oriented biochemistry. During the 1950s, both the sequence concept, and the aim of determining it, increasingly pervaded the laboratories of other biochemists and biological researchers.

1.6   Expansion and the convergence with molecular biology

The progressive completion of insulin (1951–55), and the awarding of the Nobel Prize in Chemistry to Sanger in 1958, prompted an expansion of sequence determination within protein research. During the 1950s and 1960s, an increasing number of biochemists either adopted the technique or developed alternative sequence determination methods. In either case, researchers assimilated the form of work Sanger's technique embodied: their use of sequence determination was based on the practices and instruments of analytical and synthetic chemistry.

One of the first investigators following Sanger to devise sequence determination techniques was Pehr Edman, a Swedish biochemist based at the University of Lund. During the late 1940s, Edman visited the Rockefeller Institute and interacted with members of the Laboratory of Chemistry, which had remained an active centre for protein analysis after Bergmann's death in 1944. Edman was especially interested in the sequence determination work of William Stein and Stanford Moore. By the time he returned to Sweden in 1950, Edman had decided to develop his own technique (Patridge and Blombäck, 1979).

Edman's method also started with protein cleavage, but instead of acids or enzymes he employed a chemical agent able to sever the last amino acid from the chain. By successively applying this agent and submitting the cleaved amino acid to paper chromatography and chemical analysis, Edman was able to determine sequences, unit by unit, in the correct order (Edman, 1949, 1950, pp. 288–92). The procedure was, therefore, also inspired by the 'degradation' of the protein, and the term was frequently used in Edman's papers (e.g. ibid., pp. 284 and 292).

Edman's technique – rather than Sanger's – was adopted as the basis for the automation of sequence determination. The process was initiated at the Rockefeller Institute in the late 1950s, with Stein and Moore designing an apparatus which automated amino acid analysis: the identification and quantification of the protein breakage products following separation through chromatography (Spackman, Stein and Moore, 1958).[11] Stein and Moore's strategy, described as 'subtractive', consisted of employing Edman's technique of successively cutting the terminal amino acids of the protein and then, following each cleavage, checking which amino acid was missing through the automatic analyser (Hirs, Moore and Stein, 1960, p. 633; Moore and Stein, 1972, p. 85). It was, again, based on the degradation of the molecule, and the automated activity – amino acid analysis – was exactly the same procedure Chibnall's group had been performing in their investigations of protein composition.

At the time that Stein and Moore were automating amino acid analysis, Sanger was investigating how to avoid this step in his methods.

Amino acid analysis required applying a series of reagents to the chromatogram to make the separated protein fragments and amino acids visible in the form of spots. The spots were then chemically identified and the amino acids quantified. Sanger considered this 'time consuming and tedious', and after the determination of insulin (1955) sought ways to skip the process and deduce the sequence directly from the chromatogram or alternative separation surface (Sanger, 1988, p. 12).

His main strategy, developed between the late 1950s and early 1960s, was to label the protein fragments and amino acids with radioactive substances called isotopes. When the chromatogram or alternative medium was photographed, the radioactively labelled fragments or amino acids appeared as dark bands in what Sanger described as a protein fingerprint. The sequence could, in principle, be deduced by measuring the position of the bands (Sanger, Hartley et al., 1959; Sanger and Milstein, 1961; Sanger, 1963).

Sanger had been introduced to labelling with radioactive isotopes by Christian Anfinsen, a US biochemist who had visited the Cambridge Department in 1954 (Sanger, 1988, p. 11). The use of this technique in the fields of physiology and nuclear medicine dates back to the 1930s. As part of the search for peacetime uses of radioactivity after World War II, this labelling method was applied increasingly within biology. Radioactive labelling contributed decisively to the transformation of fields such as immunology, cell biology and genetics, as well as to the interactions between them (Creager and Santesmases, 2006; Herran and Roqué, 2009; Podolsky and Tauber, 2000; Rasmussen, 1999).

In Britain, the Atomic Energy Authority developed a radiochemical centre specialising in the manufacture of radioactive tracers. This centre was later transformed into a public company called Amersham, which became the main supplier of isotopes to Sanger and other biological research groups (Gee, 2007, ch. 4). Sanger was, therefore, using a technique that bridged a variety of biological disciplines to eliminate amino acid analysis, an important chemical step in sequence determination.

Sanger had resisted radioactive labelling, suspecting that the amino acid sequence could be altered after the incorporation of the isotopes into the protein fragments. His reluctance also relates to an intense dislike of physics, which he continually attempted to avoid, having obtained poor results in the subject as an undergraduate (Sanger, 1988). This aversion helps us understand his minimal interactions with X-ray crystallographers, despite their common interest in determining protein structure.

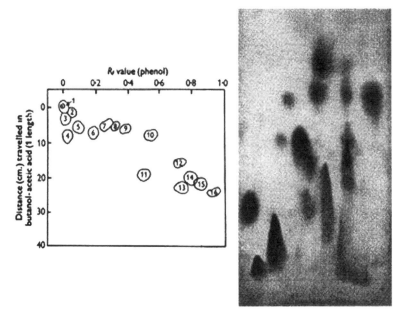

Figure 1.3 Two dimensional chromatogram (left) and protein fingerprint (right). In the former, the amino acids are first analysed with chemicals and then determined according to their relative movement. In the fingerprint, the protein fragments are labelled with a radioactive substance and can be visualised as dark bands on the separation surface after being photographed. (Sanger and Thompson, 1953b, p. 369; Ingram, 1958, p. 543. Reprinted with permission from the *Biochemical Journal* and *Biochimica et Biophysica Acta*. Copyright UK Biochemical Society and Elsevier.)

At the time that Sanger was developing his sequence determination methods, X-ray crystallographer, Dorothy Hodgkin, was studying the three-dimensional structure of insulin at the University of Oxford. She had actively sought contact with Sanger, and was disappointed by the lack of enthusiasm which he and other protein chemists appeared to show towards her results (Ferry, 1998, pp. 327–8). Sanger believed that sequence determination and crystallography were essentially different techniques, leading to related but independent structures. Although there could be contact between crystallographers and biochemists, according to Sanger this union would never achieve experimental practices. Sanger's laboratory notebooks show little interaction with Hodgkin or any other X-ray crystallographer.[12]

The development of the fingerprinting method coincided with another line of research that Sanger initiated in the late 1950s, when he attempted to deduce the functioning of insulin from its recently determined sequence (Sanger, 1988, p. 11). He did this by comparing sequences of different species and studying previously radioactively labelled active centres of enzymes (id., 1955; 1988, pp. 11–14; Sanger, Hartley et al., 1959; Sanger and Milstein, 1961; Sanger, 1963). An interest in comparison and in the functional aspects of molecular structures had been present in the Dunn Institute since the 1920s and 1930s. Baldwin, its main representative, had been Sanger's undergraduate tutor and an important influence on his career in biochemistry. Both comparison and the structure-function connection were further reinforced by Chibnall's protein analysis programme and the emergence of sequence determination as a form of work in Cambridge.[13]

Sanger was not fully satisfied with the results of any of his post-insulin projects (Sanger, 1988, p.11). The fingerprinting technique, despite being successfully used in the detection of amino acids (see Chapter 2), did not allow direct determination of sequences, while comparisons between different proteins showed that the structure-function relationship was not as direct as postulated. This apparent stagnation was one of the factors that led Sanger to initiate a new stage in his career by moving, in 1962, from the Department of Biochemistry to a new research centre opened in Cambridge. After the move, he focused his sequence determination work on RNA and later DNA. Sanger's move was decisively shaped by his increasing contact with an active group of researchers from another laboratory in Cambridge. These researchers defined themselves as representatives of a new discipline called molecular biology and were the main promoters of the new centre to which Sanger moved.

2
From Chemical Degradation to Biological Replication (1962–1977)

The impact of Sanger's determination of insulin (1955) and his subsequent Nobel Prize (1958) led to the uptake of his technique not only by biochemists, but by other biological researchers. Especially active were the members of an emerging group in the Cavendish Laboratory, close by in the University of Cambridge. The Cavendish Laboratory had been a leading centre in physics since its foundation in the late-nineteenth century, with James Clerk Maxwell, J.J. Thompson and Ernest Rutherford all producing important contributions to the structure of the atom and its electromagnetic properties (Warwick, 2003; Buchwald and Warwick, 2001; Hendry, 1984). From the early 1930s onwards, a group of Cavendish investigators were among the pioneers in the application of X-ray crystallography to proteins.

X-ray crystallography had been developed in the early twentieth century by British physicists William Henry and William Lawrence Bragg, a father and son who extensively applied the technique to minerals and rocks. Among the students of William Henry at the Royal Institution in London were William Astbury – who became a renowned protein crystallographer in Leeds – and J.D. Bernal, a charismatic and unorthodox character, who became the first leader of the Cavendish crystallographic group (Brown, 2007). Dorothy Hodgkin was educated there before moving to Oxford and focusing on the three-dimensional structure of insulin (see Chapter 1). In 1937, Bernal migrated to Birkbeck College London, having been disappointed not to obtain a senior position at Cambridge, partly due to his Marxist politics (Olby, 1992 [1974], ch. 16).

The end of World War II and the search for peacetime applications for both the technologies and researchers involved in wartime physics projects gave new impetus to the Cavendish group. The Medical Research Council (MRC) – the same government body which supported biochemistry in Cambridge – created a Unit for the Study of Molecular Structure of Biological Systems at the Cavendish Laboratory, directed by William Lawrence Bragg, who had succeeded Rutherford as Professor of Physics. This unit coexisted, at the Cavendish, with research on atomic physics and astrophysics.

Francis Crick was one of the many physicists converted to biology following World War II. Mobilised during the war and involved in military research projects, he arrived at Cambridge in the late 1940s to start a belated PhD. After some deliberation, he chose to study the three-dimensional structure of proteins and joined the Cavendish crystallographic group, where in 1951 he met a young postdoctoral visitor from the United States named James Watson. During his previous work, Watson had studied the role of genes in the phenotype – the physical and behavioural attributes of species – using viruses as model organisms.

Watson and Crick decided to address the three-dimensional structure of DNA. The substance deoxyribonucleic acid had been known since the late-nineteenth century, but research had been obscured by the dominant focus on proteins in subsequent decades. Crick himself was working on proteins for his doctoral research (Crick, 1988; Watson, 1969; de Chadarevian, 2002, Part II). However, at the time of his collaboration with Watson, a number of researchers were suggesting that DNA, rather than proteins, might be the genetic material and therefore the molecule in control of phenotypic attributes.[1]

Watson and Crick began by reviewing the investigations of the few other crystallographers and biochemists who had studied DNA. They then designed a methodology which combined crystallographic and chemical approaches, and enabled them to determine both the three-dimensional configuration of atoms in the molecule and their most plausible chemical bonds.[2] In 1953, they postulated that DNA consisted of a double helix formed by two coiled strands of the chemical nucleotides adenine, cytosine, guanine and thymine. The strands were joined by hydrogen bonds in such a manner that adenines were always attached to thymines and cytosines to guanines. This structure suggested a mechanism by which each strand could reconstruct the other. Watson and Crick explicitly linked this mechanism to the hereditary function of genes: a molecule which could replicate itself would be able to pass heritable 'information' from parent to offspring (Watson and Crick, 1953a,b).

In the following years, Watson and Crick focused their investigations on the so-called 'coding problem': how DNA, as genetic material, directed the formation of proteins – and hence metabolism and other functional processes – within a cell. Watson returned to the United States and Crick began to work with a young and enthusiastic researcher who had just joined the Cavendish Laboratory, Sydney Brenner. Brenner and Crick soon realised the potential of Sanger's technique for matching the genetic activity of DNA with the amino acid sequence of proteins. They sought to attract Sanger to their group and framed his incorporation within a broader campaign to establish a new discipline in Cambridge, one which required its own research centre.

2.1   The move to a new centre

Sanger's interactions with the Cavendish Laboratory have been reviewed in autobiographies by the researchers involved and studied in detail by Soraya de Chadarevian. These interactions began soon after the creation of the MRC unit and were particularly intense following the completion of the sequence of insulin in 1955. An emerging crystallographer, John Kendrew, attempted to attract Sanger to the Cavendish to match the three-dimensional structure of myoglobin – the protein he was studying – with its amino acid sequence (see Chapter 3). However, the strong orientation at the Cavendish towards both crystallography and physics in general, led Sanger to regard this work as remote from his own interests (de Chadarevian, 1996, p. 371). Sanger's perception became more favourable when Crick and Brenner asked him to join their investigations into the coding problem.

Crick had been aware of Sanger's insulin work since at least 1954, a year after the double helix structure was initially suggested. He had attended a public lecture series in which Sanger had explained the degradation procedure to other Cambridge-based researchers. In these lectures, Sanger described the breakage of proteins into fragments and reconstruction of the sequence from the fragments' overlaps as 'the jigsaw-puzzle method'. Sanger's work would have a profound impact on Crick's further research.

Crick used Sanger's published insulin results as evidence in the formulation of his sequence hypothesis. The hypothesis stated that, given the complexity of protein structure – as Sanger had demonstrated in insulin – it could only be generated by an equally intricate molecule, DNA. Crick postulated that the sequence of amino acids in proteins was determined by the sequence of nucleotides in DNA (de Chadarevian, 1996, pp, 375–6).

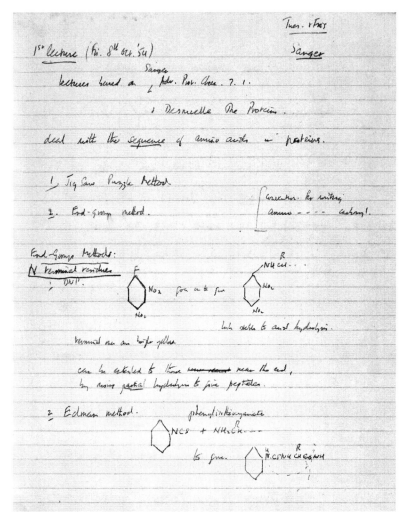

Figure 2.1 Excerpt of Crick's manuscript notes from a series of lectures delivered by Sanger which he attended in 1954, one year after the elucidation of the double helix of DNA. The 'jigsaw puzzle method' is referred to in bullet point 1. (Francis Crick's Papers, Wellcome Trust Archives, London, file PP/CRI/G/1/6. Reprinted with permission from the Wellcome Trust.)

The sequence hypothesis was published in a 1958 review paper that Crick devoted to 'protein synthesis' (Crick, 1958, pp. 138–63). Its publication coincided with a change of strategy in attempts to solve the coding problem. During the mid-1950s, Brenner and Crick, along with other researchers, had sought to mathematically deduce the rules matching protein and DNA sequences. They adapted the methods of cryptography, a technique widely used in World War II to decipher coded enemy messages (Kay, 2000, ch. 4; Strasser, 2010, pp. 631ff.; Strasser and de Chadarevian, 2011, pp. 326ff.). However, towards the end of the decade Crick and Brenner decided to apply a markedly different technique: they produced mutations in bacteriophage viruses by submitting them to radiation and studied how alterations in their very simple DNA sequences affected the formation of proteins.

De Chadarevian has shown how Brenner and Crick soon realised Sanger's techniques would be extremely useful for matching mutations in the DNA of the viruses with changes in the amino acid sequence of the resulting proteins. This led to an informal co-operation between Sanger and Brenner, in which the former determined partial sequences of proteins produced by the mutant viruses obtained by Brenner (de Chadarevian, 1996, pp. 379–82; 1999, p. 204). A more formalised and productive association was that of Sanger with Vernon Ingram, another visitor from the US investigating sickle-cell anaemia. This disease, and the structural alteration of haemoglobin which caused it, was a major line of research at the Cavendish and the object of X-ray analyses by crystallographer Max Perutz. Ingram used Sanger's fingerprinting techniques to identify the amino acid replacement in haemoglobin that triggered the disease (Ingram, 1958; see Figure 1.4).[3]

The potential of co-operation with Sanger led Crick and Brenner to intensify their efforts to attract him to the Cavendish Laboratory. They wanted Sanger to carry on applying his protein techniques to the coding problem and begin developing methods for the sequence determination of RNA and DNA. Crick and Brenner, additionally, were campaigning with Perutz and Kendrew for a new discipline and an independent institutional setting for the Cavendish biologists. Their efforts crystallised at the end of the 1950s when, impressed with the increasing prestige of the Cavendish Unit, the MRC approved the construction of a new laboratory, capable of involving other researchers. From the beginning, Sanger was perceived as an essential component of this new institution, named the Laboratory of Molecular Biology of Cambridge (LMB) (de Chadarevian, 1996, 2002).[4]

The markedly biological orientation of the new laboratory overcame Sanger's reticence and he accepted the offer. Also persuasive was the increasing favour the new molecular biology was receiving, not only in Britain, but also in France, the US and from other scientific powers (Kay, 1993; Gaudillière, 2002). The LMB had better facilities than the Department of Biochemistry and Sanger would be exempted from teaching, a duty he particularly disliked (Sanger, 1988). Sanger moved to the LMB in 1962, shortly after it opened. That same year, Watson, Crick, Perutz and Kendrew were awarded the Nobel Prize for the double helix, and the determination of the three-dimensional structures of haemoglobin and myoglobin.[5]

Prior to the move, Brenner and Crick had organised two seminar sessions in the Golden Helix – residence of the latter – where they conveyed to Sanger and other migrating members of his group the fundamentals of DNA. The speakers included Seymour Benzer, another advocate of the new discipline of molecular biology, who was visiting the Cavendish from the US. Among the topics discussed were the structure of DNA, its relation to RNA and proteins in the coding problem and, crucially, the mechanism of cell replication (de Chadarevian, 1996, p. 381).

Sanger and Crick's recollections of this period differ in their retrospective accounts. The former appears to have forgotten attending the seminars and claims to have acquired all his background on RNA and DNA from another LMB member, John Smith (Sanger, 1988, p. 14).[6] Crick, by contrast, considers the Golden Helix lectures a key factor in Sanger's move to the LMB (Crick, 1988, pp. 102–7). This apparently unimportant contradiction is representative of fundamental differences in Sanger and Crick's views of molecular biology.

2.2   Molecular biology and Sanger's professional identity

Sanger's move to a molecular biology centre and introduction into the emerging discipline did not prevent him from considering his career as existing fully within biochemistry. In his retrospective accounts, he has described his research as invariably involved in the same project – sequence determination – and surviving institutional and disciplinary shifts. When commenting on the effects of the move to the LMB, Sanger refers only to the larger space and facilities, seeing no significant changes in the nature of his work. The LMB, according to Sanger, was named Laboratory of Molecular Biology simply because 'there was already a department of biochemistry in Cambridge.' Sanger's lack of distinction between these disciplines is confirmed when he refers to Crick as a 'theoretical biochemist'.[7]

Sanger's account differs from de Chadarevian's study of the LMB foundation. For de Chadarevian, Sanger's move was a key episode in the emergence of molecular biology in Cambridge, marked by the 'alliance of protein crystallographers, molecular geneticists and protein chemists.' This convergence required 'institutional and disciplinary negotiations' – as shown by Sanger's collaboration with Crick, Ingram and Brenner – which came to fruition in the LMB, specifically built to bring these scientists together (de Chadarevian, 1996, p. 385). De Chadarevian argues, therefore, that the history of Sanger at the LMB is not one of a biochemist preserving his identity within molecular biology, but of a biochemist contributing, alongside researchers from other fields, towards the emergence of a new discipline.

In accounts of the emergence of that discipline, Crick and Brenner's opinions have prevailed over those of Sanger. The Cavendish researchers have propagated the idea that, despite Sanger having devised the techniques, it was they who envisioned 'the problems to which those techniques could be usefully applied' within the new discipline of molecular biology (quote from ibid., p. 380; Crick, 1988, pp. 102–7). This interpretation squares with Sanger's self-portrayal as a pure technologist, applying his instruments to the theories formulated by others – in this case leading and self-declared molecular biologists. It is also an account in which the collaborative effects of Crick, Brenner and Sanger are seen as unidirectional: the exportation of a set of inert techniques – sequence determination – to a dynamic new field. Crick and Brenner's personalities – eloquent, persuasive and charismatic, in contrast with Sanger's reticent nature – have reinforced the dominance of their account.

This disequilibrium illustrates a more general problem in the history of twentieth-century life sciences: the pervasiveness of retrospective accounts propagated by molecular biologists. The prestige and sociopolitical authority of this discipline has led to the proliferation of autobiographies in which prominent molecular biologists offer their interpretation of the field's development. They emphasise the importance of the *progress* in the elucidation of DNA's structure and function, and present the postulation of the double helix as the *foundational moment* of the new discipline (e.g. Crick, 1988; Brenner, 2001; Watson, 1969, 2003, 2010; Wilkins, 2005).

The accounts of molecular biologists have been influential not only on lay understandings, but also within academic studies of current science and technology. They have shaped the perception of biology after 1953 as dominated by the *revolutionary achievements* surrounding the DNA molecule (e.g. Judson, 1977, 1992; Wills, 1991; Cook-Deegan, 1994; Atkinson, Greenslade and Glasner, 2007). Other biological fields and molecules – even those, such as proteins, which were important for the development of molecular biology – are obscured by these DNA-centred stories (de Chadarevian, 2002, ch. 8).[8]

Historians have partially mitigated this effect by placing molecular biology within the context of the development of post-World War II life sciences. However, as de Chadarevian and Harmke Kamminga have argued, some scholarship on the 'molecularisation' of the life sciences has identified the emergence of a molecular vision of life with the rise of molecular biology between the 1950s and 1970s (de Chadarevian and Kamminga, eds., 1998, p. 1). This literature implicitly suggests that all the biological disciplines were transformed according to the parameters of molecular biology (Abir-Am, 1982; Kay, 1993; Olby, 1990; Fox Keller, 2000).

Sanger's early career illustrates that the phenomenon of molecularisation cannot be explained solely within the development of molecular biology. His protein sequence determination research is an example of a form of work derived from molecular analysis in biochemistry and then exported to molecular biology.[9] In this journey, Crick and Brenner's persuasion was an important factor, but not the only cause of Sanger's transition. Radioactive labelling and the belief in a straight structure-function relationship bridged Sanger's sequence determination and the interests of the Cavendish biologists. Furthermore, Ingram's use of fingerprinting techniques illustrated to Sanger the connection between protein sequence alterations and genetic diseases such as sickle-cell anaemia. Another key factor was Sanger's interest in protein synthesis, an interest which had started in the late 1950s.

2.3   Protein synthesis and the transition to nucleic acids

An analysis of Sanger's laboratory notebooks demonstrates that his interest in nucleic acids – RNA and DNA – and their connection to proteins was gradual and not prompted by a single disciplinary move. His later insulin volumes, written between 1958 and 1962, show he began his experiments on RNA sequence determination while still at the Department of Biochemistry, in parallel with those on radioactive labelling and the comparison of protein sequences. At that time, Sanger became interested in the problem of protein synthesis, the assembly of the protein once information on the amino acid sequence has been transferred from DNA to RNA. In his RNA books, written after his move to the LMB, fewer experiments on proteins are evident, but an engagement with synthesis continued to infuse Sanger's work.[10] Some of his

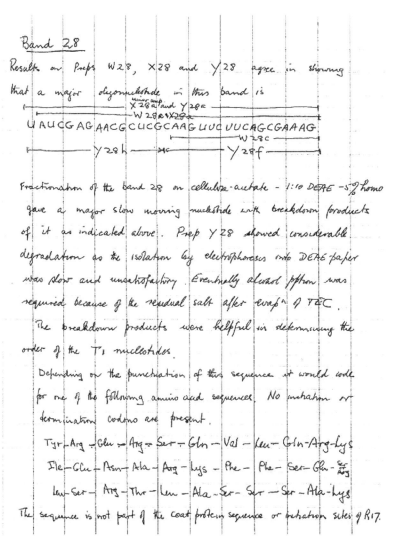

Figure 2.2 Sanger's RNA experiment during the 1960s. The RNA sequence, at the beginning of the page, is matched with the corresponding protein sequence at the bottom. (Sanger's Laboratory Notebooks, Wellcome Trust Archives, London, RNA Books, file number SA/BIO/P/1/30, experiment T28a. Reprinted with permission from the Biochemical Society.)

assistants – such as Ieuan Harris and Brian Hartley – continued work-ing on protein sequence determination, and it remained an important line of research in the LMB during the 1960s and 1970s (Sanger, 1988, pp. 102ff.; Hartley, 1970a,b).

Protein synthesis had been a concern of biochemists since the first decades of the twentieth century, particularly researchers in the Dunn Institute prior to Charles Chibnall's arrival (see Chapter 1; Florkin, 1977, vol. 32). In the mid-1950s, it also became a major interest of the emerging molecular biology, epitomising as it did the relationship between DNA, RNA and proteins (Gaudillière, 1992; Strasser, 2006b). Historian Hans-Jörg Rheinberger has argued that, between the late 1940s and 1960s, protein synthesis as an 'experimental system' gradu-ally shifted from the realm of biochemistry to become the focus of an increasing number of self-declared molecular biologists (Rheinberger, 1993; 1997, esp. ch. 10). Sanger and Crick had both adopted this exper-imental system, for differing reasons, in the years preceding the LMB foundation.

Protein synthesis for Crick was a model which captured the activity of DNA as the genetic material: a one-directional transfer of informa-tion to RNA and proteins. In the same paper where he postulated the sequence hypothesis (1958), this information transfer was described as the 'central dogma' of molecular biology (Crick, 1958, p. 153). Crick's stated dogma influenced the future investigations of both existing advo-cates of molecular biology and newcomers to the field, who increasingly considered DNA to be the *master molecule* upon which research efforts should be focused (Antonarakis and Fantini, 2006).[11] Sanger, by con-trast, had no interest in the role of genes while working on the sequence of insulin. He subsequently became concerned with protein synthesis after 1955, as a means of linking his protein work with the studies he was beginning on nucleic acids.

Sanger's engagement in protein synthesis demonstrates that his con-tacts with Crick and Brenner, as well as Ingram's experiments, were important in raising his interest in the role of RNA and DNA. However, the previous tradition of research on protein synthesis in biochemistry, together with the way Sanger introduced this concern in his notebooks, reflects a gradual incorporation rather than a dramatic shift. Sanger undertook protein synthesis for practical reasons, to give coherence to his work rather than as a disciplinary dogma. This suggests that he was not *converted* to molecular biology in a single non-critical step – such as at the Golden Helix seminars or his move to the LMB. Instead, Sanger progressively incorporated tools and strategies that were becoming common in molecular biology, in order to facilitate his transition to the study of nucleic acids.

Sanger's move to RNA and DNA sequence determination, therefore, endorses de Chadarevian and Kamminga's model of molecularisation: as a bidirectional process, relying on particular research strategies. Both historians have argued that the molecularisation of twentieth century life sciences depended not only on molecular biology, but on previous research traditions equally centred in a molecular explanation of biological phenomena (de Chadarevian and Kamminga, eds, 1998, ch. 1). In this regard, Sanger's collaborations and his relocation to the LMB transformed sequence determination from its previous biochemical identity. However, sequence determination as a form of work also affected the way in which molecular biology was conducted. Sanger's techniques had a meaning of their own, and their incorporation into research at the LMB transformed the problems to which Brenner and Crick applied them.

2.4 The development of new techniques

Sanger's first task at the LMB was to expand his experiments on RNA sequence determination. He did so with a similar method to that used in his later insulin work. The RNA was degraded with enzymes and the resulting fragments radioactively labelled and separated, so that by photographing the separation surface – also two-dimensional – the sequence could be deduced by detecting overlaps between the fragments (Sanger, 1988, pp. 14–20). Sequence determination, therefore, remained mainly within the parameters of analytical and synthetic chemistry, but incorporated an instrument – radioactive labelling – that bridged different biological fields and became an essential tool in the emerging molecular biology.

With this technique, Sanger determined the sequence of a number of short RNA molecules involved in protein synthesis – transfer and ribosomal RNAs (Sanger, Brownlee and Barrell, 1965; Brownlee, Sanger and Barrell, 1968). Sanger's work attracted increasing interest and contacts, not only at Crick and Brenner's laboratory, but also within the growing international institutions embracing the new molecular biology. From the second half of the 1960s onwards, a number of molecular biology laboratories – either new or renamed – devoted considerable efforts to determining longer RNA sequences.[12] Equally, Sanger's assistants at that time had less specifically biochemical profiles: Bart Barrell, joining the laboratory in the early 1960s, was educated to A level without a university background, and George Brownlee, appointed at the same time, came from studying a more general biological degree in Cambridge.

When Sanger attempted to apply the degradation procedure to DNA, he found it was unsuitable, due to this molecule being longer than proteins and RNA, and there not being enzymes to cleave it at specific sites in its sequence. This forced him to create an 'entirely new approach' to sequence determination, one based on duplicating rather than degrading the molecule, which he named the *copying procedure*, (Sanger, 1988, p. 20). In this approach, developed during the late 1960s and 1970s, the DNA was no longer successively broken, but replicated with an enzyme (polymerase) which, in the presence of a series of loose nucleotides, progressively added them to the DNA to be determined – the template DNA.[13] The addition followed the rules of chemical complementarity between the nucleotides in the two strands of the DNA double helix: adenines were always matched with thymines and cytosines with guanines (Sanger, 1988, pp. 20–1; 1980, p. 431; Sanger and Dowding, 1996, pp. 339–44; Fruton, 1999, pp. 412–13).

Sanger devised two methods based on this copying procedure. In both of them, he prepared four different test tube reactions with polymerase, various copies of the template DNA and loose nucleotides. He then selectively stopped the action of polymerase at a different nucleotide in each reaction. In the plus and minus method, the selective stop of polymerase was achieved by including either a bigger or smaller concentration of one of the loose nucleotides (adenine, cytosine, guanine or thymine) in each test tube. This way, polymerase would stop more frequently at the more numerous nucleotide (the plus method) or before the scarce one (the minus method) (Sanger and Coulson, 1975 pp. 441–8; Sanger, 1975, pp. 324–8). In the dideoxy method, normal nucleotides were combined with chemically modified ones (dideoxy) which were able to halt the action of polymerase. This permitted the achievement, in each test tube, of DNA fragments which respectively finished at dideoxiadenine, dideoxicytosine, dideoxiguanine and dideoxithymine (Sanger, Nicklen and Coulson, 1977, pp. 5463–7).

The outcome of both methods was a series of overlapping fragments in which the last nucleotide (adenine, cytosine, guanine or thymine) was known. If the fragments of each reaction were gathered and separated by size, the sequence could be deduced by analysing their ends – progressively bigger – on the resulting separation surface (Sanger, 1980, pp. 432–7; Wills, 1991, pp. 40–5).

The copying procedure incorporated gel electrophoresis, a different separation method from chromatography. This technique was based on

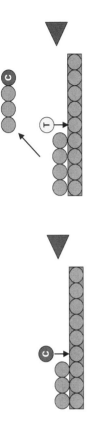

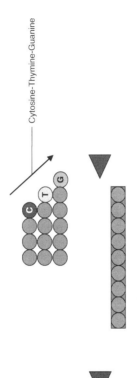

The DNA to be sequenced is duplicated with polymerase, an enzyme which progressively adds nucleotides to the chain

By different means, polymerase is selectively stopped at adenine, cytosine, guanine and thymine

Cytosine-Thymine-Guanine

By performing this operation in various DNA samples and ordering the fragments by size, the sequence can be deduced from there sulting *ladder*

Elaborated by author

Figure 2.3a Sequencing DNA (copying approach)

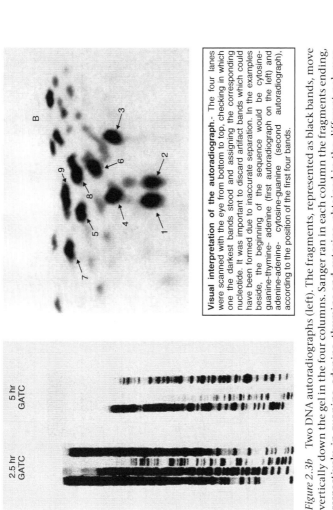

Visual interpretation of the autoradiograph.- The four lanes were scanned with the eye from bottom to top, checking in which one the darkest bands stood and assigning the corresponding nucleotide. It was important to discard artifact bands which could have been formed due to inaccurate separation. In the examples beside, the beginning of the sequence would be cytosine-guanine-thymine- adenine (first autoradiograph on the left) and adenine-adenine- cytosine-guanine (second autoradiograph), according to the position of the first four bands.

Figure 2.3b Two DNA autoradiographs (left). The fragments, represented as black bands, move vertically down the gel in the four columns. Sanger ran in each column the fragments ending, respectively, in guanine, adenine, thymine and cytosine, obtained in the different polymerase reactions. The separation was, hence, different from the one conducted during his protein and RNA work (right). (Sanger, 1980, p. 436; Jeppesen, Barrell, Sanger and Coulson, 1972, Plate 2, unnumbered page. Reprinted with permission from the Nobel Foundation and the *Biochemical Journal*. Copyright The Nobel Foundation and the Biochemical Society.)

the application of an electric charge to a porous gel on which molecular substances had been placed. The molecules were separated according to their different mobilities across the gel (Thurtle, 1998). Sanger divided the gel into four lanes and placed in each of them the DNA fragments from the four different polymerase reactions (respectively ending in adenine, cytosine, guanine and thymine). When the electric charge was applied, the fragments migrated vertically from top to bottom of the gel. Bigger fragments moved slower, smaller ones faster. The result was a different pattern of fragments in each lane (Sanger, 1988, pp. 22–4; Wills, 1991, pp. 40–5).

Sanger radioactively labelled the DNA fragments before separation, as he did in his later protein and RNA work. After applying electrophoresis, he also produced a picture of the gel, called an autoradiograph. Due to the ladder-like overlapping nature of the fragments, the DNA sequence could be deduced by scanning the autoradiograph with the eye. The researcher began at the bottom of the picture, checking for a dark band – the product of radioactivity – in the four lanes. By progressively moving up and repeating the operation, it was possible to determine the sequence (see previous page, Figure 2.3b). The visual interpretation of the autoradiograph – as well as the other steps of the copying procedure – necessitated dexterity at the laboratory bench and the learning of specialised skills. This made Sanger's laboratory at the LMB a key centre for training researchers in DNA sequence determination.[14]

2.5   The insufficiency of a technological explanation

Sanger has persistently claimed that his shift from degradation to a copying procedure was largely due to the different availability of technologies between the 1960s and 1970s. In both his autobiographical account and interviews, he has stressed the role of a series of technical instruments which, allegedly, became available at the time he developed his DNA methods. These instruments, namely polymerase and gel electrophoresis, were, according to Sanger, the crucial factors in his change of approach once he realised that the degradation procedure was unsuitable for DNA (Sanger, 1988; Lagnado, 2005).

Sanger's narrative is, to a large extent, based on his self-identification as a 'technologist', a scientist concerned with the development of methods rather than with the formulation of hypotheses.[15] However, a closer look at his transition to copying suggests this account may have been created to fit with Sanger's self-definition. The main evidence for this is Sanger's hesitation about the matter in a passage of his autobiographical paper. In it, he admits that the account may have been retrospectively constructed:

> [The] new approach to DNA sequencing was I think the best idea I have ever had (...), so I have attempted to describe its development in some detail, but on reading it through I must confess that I am by no means certain that it really did happen like that. I certainly do not remember having the idea, whereas I do remember doing some preliminary experiments and discussing it with [my assistants] Alan Coulson and John Donelson. I have a feeling that the above account [entirely technological] may have originated to some extent from my attempts to explain the method in a simple way when giving lectures, and that subsequent frequent repetition resulted in its being established as part of my 'official', but perhaps not actual, memory (Sanger, 1988, p. 22).

Sanger's account may, thus, be a simplification in light of the further success of the copying procedure. It presupposes a Eureka moment – having the copying *idea* – and a single shift in which the method was developed by adopting the available technologies. The technological explanation, therefore, squares with Crick and Brenner's view of Sanger entering into molecular biology and changing his sequence determination procedures in one simple step. However, his late 1960s and 1970s experimental records reveal a substantially different process.

Firstly, Sanger's account considers only his successful experiments. His laboratory notebooks, however, demonstrate that Sanger's failed attempts and dead ends were at least as significant as his viable procedures, which did not develop in a linear fashion. The experiments are frequently accompanied by comments such as 'where does everything go?', 'don't seem to make much sense' or 'look ghastly', referring to the chromatograms or autoradiographs.[16] These comments, and parallel alternative attempts, became especially prolific around 1966, when Sanger and his assistants directed their techniques towards DNA.

The team first produced hybrid templates of RNA and DNA, in order to facilitate the specific cleavage of the molecule. They had utilised polymerase since around 1970, but not with the intention of enabling a copying procedure. he aim had initially been to create shorter DNA fragments and then apply the degradation strategy. These polymerase-derived fragments were submitted to two- and then one-dimensional electrophoresis by John Donelson, a US biochemist who pioneered the application of gel electrophoresis at the LMB during a postdoctoral stay.[17] Nevertheless, his goal was to reconstruct the sequence from the overlaps rather than to scan it visually, as in the copying approach. Sanger and his assistants published a number of papers on small DNA sequences, as revealed by these techniques (Murray, 1970; Robertson et al., 1973; Sanger et al., 1973).

In 1973, Sanger reported in his notebooks the duplication – or 'copying' – of a DNA template 'with all four' nucleotides. Shortly afterwards, he described a 'fairly ambitious experiment' which incorporated the main features of the copying procedure. The method required significant refinements in subsequent experiments and the expressions 'plus' and 'minus approach' gradually gained significance. Also gradually, the experiment reports acquired a homogenous format, and a model DNA sequence – that of bacteriophage virus ØX-174 – was determined using the new copying approach.[18]

Sanger's copying approach emerged, therefore, not as the product of a Eureka moment, but as a gradual development. As with his transition from protein to nucleic acids, it was a consequence of practical decisions while developing his sequence determination techniques, not an abrupt shift towards molecular biology. Sanger's career, in this regard, presents the features of what F.L. Holmes has called an 'investigative pathway': a research trajectory which does not develop in a linear fashion, but shows an overall continuity (Holmes, 2006, pp. VIII and IX).[19] The important factors that propelled Sanger's investigative pathway were not technologies and their availability, but rather his reasons for combining them in the copying approach.

This is especially true since the same technologies were being simultaneously used during the 1960s in sequence determination and other structural analyses of nucleic acids. Researchers in the United States and Europe were applying polymerase and gels to RNA and DNA at the time Sanger was developing his copying procedure. This contemporary use, apart from further undermining the availability of technologies argument, demonstrates that sequence determination was becoming a widespread form of work, with multiple ramifications and approaches.[20]

Polymerase had been isolated in 1956 by biochemist Arthur Kornberg and applied to RNA and DNA sequence determination in the subsequent decade. Molecular biologists at Stanford and Cornell Universities, A.D. Kaiser and Raymond Wu, used it in the late 1960s to determine the sequence of the sticky ends of the bacteriophage λ, two short single-stranded fragments at the ends of its DNA (Wu and Kaiser, 1968; Wu and Taylor, 1971).[21] One year later, the Zurich-based researcher, Carl Weissmann, applied a variation of the enzyme to the RNA virus Qβ.

D 93   One-dimensional sequencing with restriction enzyme primed products
(D 91 (5) and D 87 ref.)

This was a fairly ambitious experiment taking restriction enzyme fragments and extending them with all four deoxys, isolating the products
on Agarose, then taking samples and extending them in the absence of
the different triphosphates. All of these were done in parallel and
the final products dissolved up in 8 µl water.

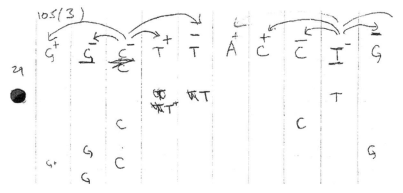

Figure 2.4   Above, Sanger's early copying experiment, conducted around 1973. Below, determination of the sequence of the bacteriophage virus ØX-174 with the plus and minus method. (Sanger's Laboratory Notebooks, Wellcome Trust, London, DNA Books, files number SA/BIO/P/1/42 and SA/BIO/P/1/43, experiments D93 and D105(3). Reprinted with permission from the Biochemical Society.)

This enzyme, RNA polymerase, was able to duplicate ribonucleic acids (Billeter et al., 1969).

Kaiser, Wu and Weissmann were, thus, creating an approach to sequence determination based on reproducing a biological process – RNA and DNA replication with polymerase – rather than degrading and reconstructing the molecules. In a retrospective account, Wu has formulated this idea as evidence of the role of his group in developing the copying approach to DNA sequence determination, usually attributed solely to Sanger (Wu, 1994). Wu and Weissmann's approaches were, however, combined with other non-biological instruments to deduce the sequences: the former used statistical analyses to measure the incorporation rates of the different nucleotides by polymerase, whereas Weissmann applied two-dimensional chemical separation (Wu and Kaiser, 1968, pp. 528–34; Wu and Taylor, 1971, pp. 496–509; Billeter et al., 1969, pp. 1085–6). In the case of Sanger, the transition to biology would be more significant.

A key element for Sanger's transition was the adoption of gel electrophoresis as the separation method for his DNA techniques. Separation by electrophoretic behaviour – movement after an electric charge – had been introduced by Tiselius in the 1930s and extensively applied to amino acids and protein fragments (Thurtle, 1998; Kay, 1988; Pederson, 1983; Gordon, 1977). In the subsequent decades, electrophoresis was developed on different media – such as paper or starch – and during the 1960s a number of researchers, largely with medical motivations, used gel electrophoresis for qualitative and quantitative analyses of nucleic acids. They did so by dividing the gel into lanes and running a different RNA or DNA molecule – previously radioactively labelled – in each lane. By comparing their different mobilities, they estimated the 'species' of nucleic acid molecule (e.g. Dingman and Peacock, 1968; Peacock and Dingman, 1968).[22]

Sanger had sporadically used electrophoresis on different media – especially paper – during the development of his protein and RNA techniques. However, it was not applied systematically in his laboratory until the first attempts to determine DNA sequences in the late 1960s. Donelson – who had been a student of Wu – was the first to apply electrophoresis at the LMB, but mainly within the degradation approach. With the advent of the copying procedure, Sanger introduced a key modification in the handling of the gel: instead of running various DNA molecules, one in each lane, he produced fragments ending in each of the four nucleotides from a single DNA template, and used the lanes to separate the nucleotides rather than the molecules. This way, the lanes were not representative of a species of DNA, but of a group of fragments ending, respectively, in adenine, cytosine, guanine and thymine.

Gel electrophoresis and polymerase were, thus, differently applied to sequence determination by Sanger, rather than introduced for the first time into this form of work. Both instruments had a recognisable tradition of use in biomedicine – including sequence determination – when Sanger incorporated them into his DNA copying procedure. This suggests that, rather than their availability, the most significant factor was their being combined innovatively at the LMB in a new approach to sequence determination.

Additionally, this approach started to be referred to as *sequencing* by Sanger and the increasing number of researchers using the DNA techniques at the LMB and abroad. The term first emerged in Sanger's laboratory notebooks in the early 1970s and was used in the paper where he described the plus and minus method, published in 1975.[23] It was then incorporated into the title of the article 'DNA sequencing with chain terminating inhibitors', where Sanger and his co-workers described the dideoxy technique (Sanger et al., 1977). The term spread among molecular biologists during the 1980s and became the dominant way of describing sequence determination. Sequencing, as well as being differently named, had significantly changed in character when compared to Sanger's previous degradation approach.

2.6   Sequencing and the move towards biology

The insufficiency of a technological explanation raises the question of what made Sanger combine polymerase and gel electrophoresis in the way that resulted in his DNA copying approach. A tentative answer is the novel institutional and disciplinary setting of the LMB, in which Sanger was immersed from 1962. If the dominance of chemical analysis and synthesis at the Department of Biochemistry had framed Sanger's degradation procedure, the new research problems, methodologies and instruments used at the LMB may have created the environment which, beyond the technologies, shaped the emergence of copying.

This environment has been analysed in detail by de Chadarevian, who defines the LMB as a 'federative structure' marked by three independent divisions. The first two had started at the Cavendish Laboratory and focused on X-ray crystallography and the emergent field of molecular genetics. Whereas the crystallography division – led by Perutz and Kendrew – maintained its interest in the three-dimensional structure of proteins after the move to the LMB, the molecular geneticists – headed by Crick and Brenner – concentrated on DNA as, according to the central dogma, the molecule in control of protein synthesis. The third division, led by Sanger, entered the laboratory in 1962 and simultaneously developed protein and nucleic acid sequencing techniques (de Chadarevian, 1996, 2002, esp. ch. 8). Sanger increasingly focused on the latter and left proteins to the other members of his growing research group.

During the early and mid-1960s – at the time of the foundation of the LMB and Sanger's relocation – molecular geneticists were shifting towards new research areas, following the resolution of the coding problem in 1961. One of these areas was cell division, pursued by Brenner in cooperation with François Jacob and other researchers at the Pasteur Institute in Paris, an increasingly important centre of molecular biology (Gaudillière, 1993; 1996).[24] The introduction of Brenner and Jacob to this problem, according to the former, led to a shift in the investigation of cell division from the perspective of biochemistry to that of molecular biology (Brenner, 2001, p. 108).

Brenner supports this claim by pointing to a change in the methods they used to investigate how DNA directed the copy of the cell by replicating itself. Shortly after the isolation of polymerase in the mid-1950s, a number of biochemists had put this enzyme and DNA into a test tube, with the aim of analysing the process. Brenner and Jacob, rather than artificially reproducing the event, sought to understand how it worked in a living organism which they had already used, the bacterium *E. coli*.[25] In their investigations, they also applied radioactive labelling and a strategy Brenner had used in his previous research on the coding problem: producing mutations in *E. coli*'s DNA and analysing their effects in the replication and cell division processes (ibid., pp. 108–15).

As these experiments were being conducted (1962–63), Sanger was developing his RNA technique through a procedure still framed in chemical degradation. His work was closely followed by Brenner, who maintained regular exchanges with Sanger and used RNA sequence determination in his cell division experiments. Brenner characterised his *E. coli* mutants by seeking alterations in the sequences of the messenger RNA produced by the mutated genes.

Sanger's group grew significantly throughout the 1960s and its membership gradually changed from researchers educated in pure biochemistry to those from a more general biological background. This led to an increase in the influence of the group's members on the techniques developed, as they incorporated practices from a wider variety of fields. Alan Coulson arrived at the laboratory in 1967, having completed a degree in Applied Biology at Cambridge. Coulson, with Donelson, was within the group of researchers more directly involved in the early development of DNA methods. His appointment was more permanent than that of international visitors, leading Coulson to become one of Sanger's closest assistants. Coulson stayed in the laboratory until the emergence of both the plus and minus method, and the dideoxy techniques, and was co-author of the 1975 and 1977 papers.[26]

Sanger's laboratory notebooks demonstrate that Coulson's role in the 'fairly ambitious' experiments which triggered the copying approach was decisive. Coulson was involved in the first attempt to use polymerase with the four DNA nucleotides, its further refinement and the emergence of the plus and minus method between 1973 and 1975. The further development of these experiments coincided with Donelson's departure from the group, following his attempt to use polymerase and gels within the degradation procedure.[27]

When reflecting retrospectively on the copying approach, Coulson believed its success and pervasiveness were due to the fact that this approach, when applied to DNA sequencing, reproduced the natural functioning of the cell.[28] As in cell division, both the plus and minus and dideoxy methods were based on the duplication of the DNA molecule, triggered by the action of polymerase. Coulson's statement suggests that, during the mid-1970s, sequence determination began to be modelled on the process of DNA replication. In so doing, it partially abandoned its biochemical identity and entered into the experimental strategies and concerns of the LMB molecular biologists.

Sanger's immersion in molecular biology had started before his move to the LMB and gradually grew in intensity and scope. Shortly after his arrival in 1962, he began using the *Journal of Molecular Biology* as the main publishing medium for his papers, rather than the *Biochemical Journal* where he had published most of his protein work. The *Journal of Molecular Biology* had been founded by Kendrew (the LMB crystallographer) and became the reference publication for researchers in the new field. Sanger published his RNA and DNA techniques here, and the number of references in his articles to molecular biology problems increased. A 1975 lecture he delivered shortly after the publication of the plus and minus method opened with the following statement:

> DNA, the chemical component of the gene, plays a central role in biology and contains the whole information for the development of an organism, coded in the form of sequences of the four nucleotide residues. The lecture describes the development and application of some methods that can be employed to deduce sequences in these very large molecules. (Sanger, 1975, p. 317)

Sanger has argued that he adopted this language simply because it was the 'gospel' of molecular biology. He claims to have referred to the concepts and concerns of his new home institution, without their having affected his sequencing techniques. For his experiments, Sanger has stated 'it did not really matter' how he was 'thinking on DNA'. The really important issue for him, as a technologist, was the development of sequencing methods through the adoption of available instruments.[29]

It is probably true that Sanger was unaware of his transition to molecular biology. However, a detailed analysis of his copying procedure shows that it was decisively shaped by his relocation. Firstly, in both the plus and minus and dideoxy methods, Sanger used part of the DNA nucleotides and the polymerase of *E. coli* – the same model organism as Brenner and Jacob (Sanger and Coulson, 1975, p. 443; Sanger et al., 1977, p. 5463). Secondly, instead of chemically intervening in the DNA template by breaking it with acids or enzymes, Sanger left polymerase to act on the molecule and to reproduce the natural process of DNA replication. Thirdly, as Brenner and Jacob had done with the *E. coli* mutations, Sanger directed the replication process towards his desired outcome by selectively stopping the action of polymerase on the DNA template. Fourthly, he introduced radioactive labelling and electrophoresis for the purpose of visualising the sequences, without having to chemically analyse them. Fifthly, when enzymes to specifically cleave DNA – restriction enzymes – became available in the mid-1970s, Sanger did not modify the copying approach as the basis of his sequencing strategy. And, finally, around 1975 Sanger began applying his methods to the DNA of a living organism, the virus ØX-174, the same type (bacteriophage) as those Brenner and Crick had used in their coding problem experiments in the late 1950s.

Sanger was, therefore, incorporating into sequencing the methodology, instruments and strategies that molecular biologists had used in the study of cell division. The development of his DNA copying approach was triggered by a transition from the practices of protein chemistry to those of molecular biology, embodied in the Molecular Genetics Division of the LMB, his new home institution.[30] Although Sanger had kept within the parameters of analytical and synthetic chemistry during the 1960s, contact with Brenner's research questions, and assistants such as Coulson, led him in the following decade to introduce replication as the basis of DNA sequencing. This made DNA sequencing different from his previous techniques, those modelled on the opposite process of chemical degradation.

Sanger's sequencing methods, however, still maintained some of their original biochemical identity. The DNA template, firstly, was not replicated *in vivo*, but *in vitro*, and the development of the dideoxy method required a large amount of analytical and synthetic chemistry to obtain the modified nucleotides. Secondly, unless the DNA template was unusually short, various samples of the molecule needed to be cleaved with restriction enzymes in overlapping fragments. These fragments were then sequenced independently through the copying procedure and the sequence, finally, reconstructed by detecting the overlaps.

The combination of biochemistry and molecular biology in Sanger's DNA methods shows how researchers routinely overcome disciplinary boundaries by developing forms of work at the laboratory bench. By focusing on the development of these forms of work, the historian is enabled to see disciplinary transitions in a new light. Academic disciplines are, to a large extent, strategic constructions of researchers, political institutions and funding agencies – as seen in the foundation of the LMB. They affect the forms of work of researchers working within them, but their shaping force is not unlimited. Sanger's sequencing was, thus, shaped by the emergence of molecular biology in Cambridge but, at the same time, developed uninterruptedly as a form of work derived from biochemistry. Other research groups, devising sequencing methods contemporaneously to Sanger within other disciplinary settings, combined biochemistry and molecular biology in a different fashion.

2.7   Rival attempts and the idea of 'elegance'

At the same time as Sanger was developing his plus and minus and dideoxy methods, an alternative approach to DNA sequencing was being invented in the US by Walter Gilbert, with the help of Allan Maxam. Gilbert was a molecular biologist based in Harvard University, the same institution to which Watson had moved after the elucidation of the double helix at the Cavendish Laboratory. During the early 1960s, Gilbert and Watson worked with other researchers on the coding problem. Gilbert had led his own research group in Harvard since 1964. Despite his orientation towards molecular biology, he had always showed a strong interest in chemistry. The development of his sequencing techniques was inspired by Soviet biochemist, Andrei Mirzabekov, who visited Gilbert's laboratory in 1975 (Gilbert, 1980, p. 409; Cook-Deegan, 1994, pp. 61–4; Sutcliffe, 1995).

This meant that Gilbert's technique combined chemistry and biology in different proportions than at the LMB. Gilbert, like Sanger, used radioactive labelling and gel electrophoresis to produce a band pattern which did not need to be chemically analysed. However, instead of applying polymerase, he submitted the DNA sample to chemicals which specifically cleaved the molecule at each nucleotide (Wills, 1991, pp. 45–7). The idea of using such chemicals was suggested by Mirzabekov during his 1975 visit (Cook-Deegan, 1994, p. 64; Gilbert, 1980, pp. 409ff.). By successively cleaving four groups of identical DNA templates at adenine, cytosine, guanine and thymine, Gilbert obtained the same overlapping fragments which Sanger was achieving with polymerase. The rest of the process was the same as that of the LMB: the fragments were separated through gel electrophoresis and the sequence deduced from the band pattern formed in the autoradiograph (Maxam and Gilbert, 1977; Sutcliffe, 1995).

Gilbert's technique was available the same year as Sanger's plus and minus method (1975) and published in the same journal issue as the dideoxy paper (1977). During the second half of the 1970s, both sequencing approaches were tested in a number of laboratories. The assessment was initially favourable to Gilbert, whose technique spread more rapidly, partly due to Sanger's method only being applicable to single-stranded DNA.[31] Moreover, Maxam and Gilbert published a further and more detailed description of their technique, substantially easing its adoption into other laboratories (Maxam and Gilbert, 1980). However, towards the mid-1980s, Sanger's method became the standard sequencing procedure, relegating Gilbert's to a secondary role.

Neither researchers nor scholars have been able to explain this preference on technical grounds alone. Whereas Sanger talks about being lucky, Coulson refers to his boss's method as 'more elegant'.[32] By elegant, he means the procedure was more easily applicable to a laboratory during the late 1970s and 1980s. The fact that the LMB was a molecular biology laboratory suggests that Coulson's idea of elegance refers to the capacity of Sanger's sequencing to adapt to the practices and working standards of this discipline, wherein the bulk of its early users were found.

This leads, again, to Coulson's remark that Sanger's sequencing reproduced the natural functioning of the cell. Researchers in molecular biology were used to these kinds of procedures, which incorporated the natural workings of DNA into their experiments. Brenner had used the mechanisms of DNA expression and replication, with Crick and then Jacob, in his attempts to solve, respectively, the coding problem and cell division during the late 1950s and the 1960s. This meant that, for these investigators, a technique such as Sanger's, which employed polymerase as a duplication agent, appeared more natural and elegant than a method which degraded DNA with chemicals. Gilbert's technique was therefore perceived as compromising the security and day-to-day routines of a laboratory. Molecular biologists, unsurprisingly, referred and still refer to it as 'the chemical method' of sequencing DNA (e.g. Garesse and Arribas, 1987).

Sanger's copying approach was, thus, preferred for its broader engagement with the practices of molecular biology – practices inspired by the *natural* workings of DNA.[33] His method was gradually adopted in every laboratory and had an increasing impact on key areas of biological research. An indirect consequence of this – and of the preference for Pehr Edman's technique in protein sequence determination (see Chapter 1) – was that Sanger's sequencing came to be identified with DNA and placed in the retrospectively constructed history of progress around this molecule. This appropriation constitutes an example of the colonising influence that molecular biology has exerted over other biomedical fields during the second half of the twentieth century (Abir-Am, 1982, 1992).

The account of Sanger's career I have offered challenges its appropriation by molecular biologists. Sequence determination originated in a purely biochemical environment during the 1940s, a decade before DNA was identified as the genetic material. Its incorporation into molecular biology occurred only when it had been consolidated as a form of work engaged in the chemical degradation of proteins. In this transition, sequence determination was affected by the practices, strategies and problems of its new disciplinary setting but, conversely, it transformed the conduct of molecular biology. Following the spread and generalisation of dideoxy sequencing in the late 1970s, molecular biologists increasingly directed their attention towards the *sequence of DNA information*, using computational technologies in that endeavour.

Part II

Mechanisation – 1: Computing and the Automation of Sequence Reconstruction (1962–1987)

> Through the 1970s, a small group of individuals began to realize that computers and sequence information were a natural marriage. Bride and groom struggled to overcome vast cultural differences. Computer scientists and molecular biologists traced their lineage through different tribes, with vastly different norms, and only a few hardy souls could converse in both languages and command respect in both communities. The database that stored sequence data became their meeting ground.
>
> Cook-Deegan, 1994, p. 285.

The use of computers and references to the sequence of information in DNA were not new to molecular biology when Sanger and Gilbert's techniques were consolidated in the late 1970s and 1980s. Historian and philosopher of biology Lily Kay has shown how an understanding of the activity of genes as information transfer dates back to the late 1940s, when there were increasing interactions between cybernetics, information theory and biological research. Cybernetics and information theory were those branches of engineering concerned with the circulation of electric signals, a key issue in the transmission of encoded messages, characteristic of military campaigns during World War II. After the war, biologists adopted informational models to conceptualise the hereditary message that genes were believed to convey (Kay, 2000, ch. 3). Therefore, during the 1950s and 1960s, the functioning of an organism was increasingly modelled on the computer (Fox Keller, 1995, ch. 3; id., 2000) and computers were applied not only to genetics, but also to other branches of the life sciences, such as X-ray crystallography, physiology, neurology and evolutionary biology (de Chadarevian, 2002, ch. 4; November, 2006, in press; Hagen, 1999, 2001, 2010; Suárez-Díaz, 2007, 2009, 2010; Strasser, 2010, 2011).

What was novel in DNA sequencing was not the conceptualisation of genetic information, but the way in which biologists – especially molecular biologists – began using this concept and applying computational technologies to it. In 1953, after the elucidation of the double helix, James Watson and Francis Crick published a sequel paper in which they considered it 'likely that the precise sequence of bases [in DNA] is the code which carries the genetical information'. Five years later, Crick postulated the central dogma of molecular biology, according to which the nucleotide sequence of DNA determined the amino acid sequence of proteins through a one-directional transfer of information (Watson and Crick, 1953b, p. 965; Crick, 1958, pp. 138–63; see Chapter 2).

Despite both papers defining DNA as a sequence, the further research agenda that Watson and Crick promoted for molecular biology was largely inspired by the notions of code and one-directional transfer of information.[1] These notions constituted a conceptual framework which, during the late 1950s and 1960s, shaped investigations into the processes of cell division, gene regulation and protein synthesis (Kay, 2000, chs 4–6). DNA was not considered information because of its sequential nature, but because its genetic code could determine other entities – direct the synthesis of proteins or its own replication in an identical DNA molecule. In this context, the transfer of genetic information mimicked the functioning of the most widely known computers of the time, central mainframe apparatuses which, having conducted a series of logical operations, were able to transform input data introduced by the user into a defined output (Fox Keller, 1995).

During the first half of the 1970s, a number of molecular biologists, including a close associate of Crick, Sydney Brenner, focused their investigations on a 'program' which would allow the 'computation' of the genetic mechanisms of development and behaviour. Taking model organisms with behavioural or developmental alterations, they attempted to determine the genetic mutations involved, much as a computer might logically deduce an outcome from multiple interrelated variables (quote from Brenner, 1973, p. 271; Hotta and Benzer, 1972; de Chadarevian, 1998b; García-Sancho, 2012).

In 1975, when presenting his plus and minus sequencing technique, Sanger stated that DNA contained 'the whole information for the development of an organism, coded in the form of sequences'. This definition differed from Watson and Crick's, in that Sanger's emphasis was on the information, rather than the code. The importance of the DNA sequence lay in the information it coded rather than on the code which carried such information. Moreover, Sanger defined his techniques as 'methods' that could be employed to 'deduce sequences' in those 'very large molecules' (Sanger, 1975, p. 317).

Sanger's definition of genetic information suggests that the incorporation of sequencing as a form of work affected not only the technical, but also the conceptual repertoire of molecular biology: his techniques introduced a means to obtain DNA sequences and, therefore, multiplied the scope of the concept of genetic information. Throughout the late 1970s and 1980s, molecular biologists were increasingly inclined to determine the information contained in the DNA sequences by applying Sanger's techniques, rather than by computing it from behavioural or developmental mutations. This transition mirrored that triggered by protein sequence determination among biochemists, who – during the 1950s – gradually abandoned the periodicity hypothesis and embraced chemical analysis (see Chapter 1). DNA sequencing transformed genetic information into an aim, and the computer and the database into privileged instruments for gathering, storing and analysing sequence data. This use of computational technologies contrasted with the previous modelling of the genetic program on the functioning of mainframe apparatuses (García-Sancho, 2012).[2]

Part II of this book will explore the consequences of the incorporation of computers and databases into the form of work characteristic of sequencing. This will connect my proposed history of sequencing with the historiography of computing, and especially with the changing identities of the computer and the notion of computation throughout the second half of the twentieth century (Campbell-Kelly and Aspray, 1996; Ceruzzi, 1998; Agar, 2003). The increasing interactions between computing and various biological fields, from the 1940s through to the 1980s (Lenoir, 1999; Hagen, 1999, 2001, 2010; Dietrich, 1998), introduced to the biological laboratory a variety of previously alien practices and professionals. In the next two chapters, I will argue that the incorporation of computers and databases into sequencing contributed to the gradual acceptance of these practices and professionals within the parameters of biological research. Their acceptance was evidenced in sequencing software, databases designed for increasingly individualised computers and, later, the emerging field of biocomputing. The incorporation of these technologies demonstrates that the interactions between biology and computing have been bidirectional, as suggested in previous scholarship (de Chadarevian, 2002, ch. 4; November, 2006; Chow-White and García-Sancho, 2012).

I will address two partially overlapping cases of the introduction of computing technologies into DNA sequencing: the development of sequencing software by Sanger's group in Cambridge (UK), and the emergence of the first international centralised DNA sequence database at the European Molecular Biology Laboratory in Heidelberg (EMBL). Both of these cases followed the invention of sequencing techniques in the mid to late 1970s and spanned the following decade. As in the previous two chapters, I have conducted interviews with the designers of these computer tools – different individuals from those who invented and used the sequencing methods – and have critically assessed their retrospective accounts in light of published and unpublished sources – including PhD theses, grey literature and both catalogued and uncatalogued archives. The archives include (1) the computing records of the Laboratory of Molecular Biology of Cambridge (UK); (2) a personal collection belonging to Graham Cameron, co-designer of the DNA sequence database and current Joint Director of the European Bioinformatics Institute and (3) the Papers and Correspondence of Sir John Kendrew, the first Director-General of the EMBL. This latter collection is available at the Bodleian Library in Oxford (UK).

Analysis of these sources will allow me to show parallelisms and interactions between the practices of computer-assisted sequence analysis and other emerging computer applications, such as operating systems and word processors. These applications mechanised some of the steps of sequencing and created a division – at times hierarchical – between the biological and chemical practices of this form of work, and those derived from computing.

A number of computerised sequence analysis practices had originated in the previous protein techniques, which had incorporated computers and databases through the 1960s and 1970s. Historians Bruno Strasser and Edna Suárez have investigated these practices in detail, in the context of the emergence of protein sequences as research objects.[3] Protein sequences, and the possibility of determining them, had a decisive impact on the configuration of molecular evolution as an independent field, differentiated from others within evolutionary biology (Strasser, 2010; Suárez-Díaz, 2007, 2009, 2010; Suárez-Díaz and Anaya-Muñoz,

2008; see also Hagen, 1999, 2001, 2010; Dietrich, 1998; Sommer, 2008). My research will extend this historiography and frame it within the development of sequencing as a form of work which allowed the circulation of practices of data computation through different biological disciplines. Due to the appropriation of sequencing by increasingly DNA-focused molecular biologists, this circulation was not always straightforward.

3
Sequencing Software and the Shift in the Practice of Computation

Sanger decided to systematically employ computers in the laboratory during the mid-1970s, when his group was applying the plus and minus method to the DNA of the bacteriophage virus ØX-174. This decision may now seem unproblematic, given the size of the molecule: with over 5000 nucleotides, it was considerably larger than those sequenced previously. However, when placed within the broader scope of the interactions between biology and computing, and their application to biological sequences in particular, one may wonder why Sanger had not incorporated computers earlier. Computers had been widely used at the Cavendish Laboratory in the 1950s and then at the Laboratory of Molecular Biology (LMB), to which Sanger migrated in 1962. Beyond Cambridge, computer programs were designed to assist in protein sequence determination during the 1960s and a considerable number of researchers in Sanger's group continued working on proteins after their move to the LMB.

Sanger has admitted retrospectively that he was 'rather reluctant' to use computers, since they might 'take some of the pleasure' he got from 'looking through the sequences and seeing what could be made of them' (Sanger and Dowding, 1996, pp. 344–5).[1] This claim suggests computers were only introduced into Sanger's laboratory when the volume of sequence data was so large it was impossible to handle it manually. Nevertheless, complementary explanations may be found if *sequencing* and *computing* are analysed as historical categories. When the genealogies of the practices behind these forms of work are taken into account, the computer expertise at hand in Cambridge between the 1950s and 1970s, together with the peculiarities of the sequence determination programs, also appear to be relevant factors.

3.1   Reading-off, puzzle-solving and the first computer program

The early to mid-1970s shift from chemical degradation to biological copying resulted in different outcomes in Sanger's techniques: a one-dimensional autoradiograph divided into four lanes in DNA sequencing; a two-dimensional separation surface for proteins and RNA. The nature of the depicted bands or spots also varied: whereas in the DNA techniques each band represented a single nucleotide, in the protein and RNA methods the spots corresponded to overlapping peptides or oligonucleotides – that is, small sequences. The DNA sequences could, thus, be directly deduced by scanning the autoradiograph with the eye and determining in which lane each spot was located. The protein and RNA sequences, by contrast, needed to be reconstructed by determining the peptides or oligonucleotides in each spot and then detecting their overlaps (see Figure 2.3b). In this sense, Sanger referred informally to protein and RNA sequence determination as the jigsaw-puzzle approach, and to the interpretation of autoradiographs as the 'read-off' of DNA sequences (Sanger, 1975, p. 443; Sanger, Nicklen and Coulson, 1977, p. 5463).[2]

This reading off practice allowed the sequence to be recorded as the autoradiograph was being scanned, either by an individual or, more often, a pair of researchers, one determining the nucleotides from the spots, the other copying the data into a notebook. However, given the limitations of this technique – autoradiographs did not normally contain more than 80 nucleotides – the DNA had to be isolated in various samples, cut into overlapping fragments which were sequenced separately, and then reassembled into a whole sequence through detection of the overlaps. Having read off the autoradiographs, it was then necessary to undertake the jigsaw-puzzle approach.

Computers were introduced into Sanger's laboratory with the aim of both automating the assemblage and avoiding the errors which frequently occurred 'just in copying out sequences' (Sanger and Dowding, 1996, pp. 344). Given the lack of computer expertise in the group, Michael Smith, a British-Canadian visitor to the laboratory and future Nobel Prize laureate, offered to contact his brother-in-law, Duncan McCallum, who worked in the management division of the chemical multinational Ciba-Geigy. McCallum routinely used computers to process administrative data generated by the company. Building on this expertise, he designed a program which allowed sequences of up to 60 nucleotides to be transcribed onto punched cards – then the usual input unit of mainframe computers (see Figure 0.3). The punched cards were submitted to a remote computing centre, where a team of main-frame operators produced a printed version of the whole sequence. The program also enabled data to be edited or translated into protein sequences, and could isolate and search for certain regions of the DNA, such as the fragments where the enzymes used to cleave the molecule acted on (McCallum and Smith, 1977).

Sanger's group used this program to compile, edit and store the sequence of ØX-174 – completed in 1977, initially with the plus and minus method, later with the dideoxy technique. Due to the lack of specificity of the program, however, the final sequence printout was an unsuitable format for publication. This led Sanger's secretary, Margaret

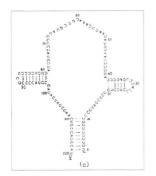

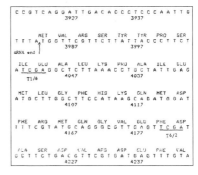

Figure 3.1 Sequence of insulin as published by Sanger in 1955 (above), together with that of 5s ribosomal RNA (below left) and ØX-174 DNA (below right). Whereas the first two preserve some of the original shape of the molecules, the DNA sequence – transcribed with a typewriter in the late 1970s – stands as a linear succession of characters. (Ryle et al., 1955, Table 1; Brownlee, Sanger and Barrell, 1968, p. 409; Sanger et al., 1978, Figure 4). Reprinted with permission from the *Biochemical Journal* and *Journal of Molecular Biology*. Copyright Biochemical Society and Elsevier.

Dowding, to manually transcribe the sequence, in the form of successive characters and consecutive lines, using a typewriter (Hutchison III, 2007, p. 6230). The published sequence of ØX-174 – based on Dowding's typed draft – contrasted sharply with that of insulin, which had appeared in 1955. In the latter, Sanger had preserved the spatial configuration of the protein's two chains, together with the disulphide bridges between them. Equally, Sanger's RNA sequences, as well as those determined by other research groups, had also maintained some of the molecule's original shape. None of these features about the molecule's shape were present in the sequence of ØX-174. Information about sequences – gathered with data-processing technologies increasingly distant from biology – was thus beginning to be detached from the chemical structure of the template molecules.

3.2   The crystallographers and the difference between sequence and structure

Having finished ØX-174, Sanger decided to carry on sequencing viruses. In contrast to his protein work, he avoided experiments that would link the sequence to its biological function, and applied the techniques to larger DNAs, such as those of bacteriophages G4 and λ. The size of these molecules, together with Smith's return to Canada, led Sanger to seek internal collaboration for computing tasks. Despite the higher sequence length of the dideoxy method – 200 to 300 nucleotides – G4 and λ yielded a number of DNA fragments that could not be assembled manually; therefore sequence assemblage was transformed into a problem that Sanger's group needed to specifically tackle.

In 1976, just as the sequencing of ØX-174 was reaching an end, Sanger sent his assistant Bart Barrell to the neighbouring Structural Studies Division in search of the necessary expertise (Sanger and Dowding, 1996, p. 344). Members of this division were X-ray crystallographers who had been using computers since the early 1950s, first at the Cavendish Laboratory and then at the LMB. They designed programs to perform the mathematical computations needed to reconstruct the three-dimensional structure of proteins following X-ray analysis. Sanger's group had seldom interacted with crystallographers and focused their contacts on Brenner, Crick and other members of the LMB Molecular Genetics Division. As the experimental strategies of this division marked the transition of Sanger's techniques from chemistry to biology during the early to mid-1970s (see Chapter 2), the incorporation of computing into sequencing derived from adaptation of the programming practices of the LMB crystallographers.

Soraya de Chadarevian has shown how the early use of computers by Cambridge crystallographers was the result of a collaboration between John Kendrew and the University's Mathematical Laboratory. In 1949, this laboratory designed a mainframe apparatus to assist other researchers at the University, titled EDSAC. The cooperation between Kendrew and the EDSAC designers was mutually beneficial, according to de Chadarevian, as computer engineers were 'keen to improve the performance of their machines' and 'computers increasingly shaped the direction and pace of work in protein crystallography' (de Chadarevian, 2002, p. 99). In 1952, Kendrew and Cambridge mathematician John Bennett designed a suite of programs which could perform the arithmetical operations 'of addition and subtraction' in one and a half milliseconds, and 'multiplication in six milliseconds' (Bennett and Kendrew, 1952, p. 109).

These and other programs proved essential in Kendrew and Max Perutz's work on the three-dimensional structure of myoglobin and haemoglobin, for which determination these Cavendish crystallographers were awarded the Nobel Prize in Chemistry in 1962, the year they moved to the LMB. Kendrew, Perutz and other crystallographers submitted a crystallised form of the protein to an X-ray beam which, after crossing the crystal, hit a photographic plate. The result of this was a diffraction pattern – a number of black spots on the plate which reflected how much each X-ray had deviated in crossing the crystal. The spot pattern was then submitted to a series of calculations, mainly Fourier syntheses and Patterson projections, in order to determine the electron density of the protein molecule. This latter magnitude enabled the position of the atoms in the protein to be modelled and, therefore, its three-dimensional structure to be investigated. Computers were introduced to solve in a manageable time – between hours and days – the tens of thousands of required calculations. They continued to be used by Kendrew and Perutz's group following their move to the LMB, and became a distinguishing feature of the Structural Studies Division (de Chadarevian, 2002, pp. 111–32).

The first LMB crystallographer that Barrell approached was Michael Levitt, who between the late 1960s and mid-1970s had written programs that used the sequences of proteins and RNAs to predict their three-dimensional folding (Levitt, 1969; Levitt and Warshel, 1975). At that time, Levitt 'felt that structure was just so much more interesting than sequence' and decided not to cooperate with Sanger (Levitt, 2001, p. 393). Levitt's rejection demonstrates that in the mid-1970s, shortly after the invention of Sanger and Gilbert's techniques, the new goal of determining DNA sequences had not universally permeated the biological mindset. Levitt and other crystallographers felt more comfortable with, and excited by, the use of mainframe apparatuses to calculate protein structures rather than process sequence information. However, the use of computers in biology was changing and this change was to a large extent fostered by a group of researchers, outside the LMB, who were designing programs for protein and RNA sequence determination.

3.3   Sequence determination and the changing use of computers

The use of the computer by Cambridge crystallographers was framed within the early identity of mainframes as mathematical and logical apparatuses. This identity was derived from the first electronic computers, developed during the 1940s and modelled on the mechanical calculators that had been in use since the mid-nineteenth century. Electronic computers were conceptualised and designed as apparatuses which were furnished with data and, by running a punched card program, applied a series of transformation rules to the data at high speed. In the case of Kendrew's group, the data input were the X-ray diffraction patterns; the transformation rules the arithmetic operations of addition, subtraction and multiplication. Following the operation, a printed output was produced showing the results of the transformation of the data input according to the program's logical rules – in this case, the calculation of the electron density coordinates. This kind of mainframe computer was used for various military operations during World War II, such as the breaking of enemy codes or the calculation of ballistic trajectories. After the war, general-purpose mainframes with stored programs emerged, among them the EDSAC in Cambridge (Campbell-Kelly and Aspray, 1996; Ceruzzi, 1998; Agar, 2003; see also Kay, 2000, ch. 3).

Following Kendrew's move to the LMB, this logical and mathematical use of the computer inspired the neighbouring Molecular Genetics Division. During the 1960s, after his work on the genetic code, Brenner proposed the nematode worm *C. elegans* as a model organism to explore the genetic causes of development and behaviour (Brenner, 1963a,b). The first papers on the worm were written in the early 1970s, and in them Brenner postulated a 'genetic programme' governed by a 'logical structure'. This programme would allow him to determine the genes involved in certain developmental and behavioural features of *C. elegans*. Having induced a series of mutations on the worm with radiation, he attempted to match the observed effects with the mutated genes (id., 1973, p. 269). Brenner was, therefore, conceptualising the nematode as a mainframe computer and using the structure of its genetic program to logically deduce certain inputs – the genes he had mutated – from their outputs – the observed behavioural attributes.[3] This experimental approach squared with the dominant concept of genetic information among the LMB molecular biologists at the time: a set of instructions transmitted from DNA to proteins and reflected in certain behavioural or developmental consequences.

At the time of Brenner's experiments, however, the use of computers and the notion of computation were changing both within and beyond biological research. The historiography of computer sciences reflects how, following World War II, an increasingly powerful hardware industry, together with emerging software companies, gradually transformed the computer into 'an electronic data processing machine rather than a mathematical instrument' (quote from Campbell-Kelly and Aspray, 1996, p. 105; see also Ceruzzi, 1998, chs 4–6; Agar, 2003, chs 5–8). During the war, computing technologies had been used in the field of operations research to organise military campaigns. On top of basic calculations, these instruments performed cross-analyses of multiple variables – such as enemy strengths, weather conditions or potential losses – in order to optimise operation strategies and the mobilisation of resources. The general-purpose computers and stored programs designed after 1945 enhanced this cross-analysis of different data.

The spread of use of the computer among large corporations between the 1950s and 1960s led cross-analysis to become one of the main uses of this technology. Insurance companies, travel agencies and government departments invested in or rented external mainframes to help in the management of records. In this new context, storage, indexing and the retrieval of data became as important as calculation to computer users. Data management duties were increasingly conducted by programmers using different programming languages, as pioneered by FORTRAN in 1953 (Haigh, 2001; Kline, 2006; Sammet, 1969, pp. 93–5; Backus, 1978; Griswold, 1978). McCallum, the designer of the first sequencing program for Sanger's group, was an example of these management experts at the chemical multinational Ciba-Geigy.

The early programming languages incorporated a number of algorithms – programming commands – designed to perform operations with strings of characters. These algorithms assisted in the composition and editing of the program by allowing searches, replacements or spellchecks of programming instructions. They were translated into machine language by compilers installed within the mainframes. The compilers analysed – in computing jargon, parsed – the syntax and grammar of the programming language and transformed it into digital information the computer could understand. As these languages developed, algorithms to search and retrieve more complex programming commands which had been previously stored and indexed within the mainframes became available. These algorithms were input into time-sharing mainframes, pooled by groups of programmers at the company or computer centre. Having completed their duties, the programmers stored and distributed the programs in punched cards and later paper or magnetic tapes (Haigh, 2006a, pp. 12–15).

One of the first uses of string manipulation algorithms beyond management was in protein sequence determination. During the 1960s, with the spread of Sanger and Pehr Edman's methods (see Chapter 1), computer programs designed to assist in the operation of sequence determination techniques began to emerge. They incorporated algorithms that could detect repetitive amino acid patterns between protein fragments, in order to assemble them into a single sequence, or between different proteins with the aim of comparing them.

Strasser has investigated the work of Margaret Dayhoff and Robert Ledley at the US National Biomedical Research Foundation (NBRF). Having originally trained as a dentist during the 1950s, Ledley worked on problems relating to operations research and the programming of mainframe computers at both the US National Bureau of Standards and George Washington University. He endorsed the controversial proposal of computer-assisted medical diagnosis and applied symbolic logic to the preparation of searchable and computerised bibliographical indices, some of which he offered to the emerging National Library of Medicine (Strasser, 2010, pp. 636ff.).[4] At the request of the US Air Force, Ledley completed an extensive survey on the possible uses of computers within biology and medicine, in which he included protein sequence determination (Ledley, 1965, pp. 371ff.). In 1960, he created the NBRF as a private charity with the explicit aim of exploring 'the possible uses of electronic computers in biomedical research' (Strasser, 2006, p. 110).

Margaret Dayhoff held a PhD in quantum physics and worked at the Watson IBM Computing Laboratory using mainframes to calculate the resonance energies in small molecules. She joined the NBRF shortly after its foundation and together with Ledley, designed the first programs to reassemble protein sequences from overlapping fragments between 1962 and 1964. The programs were written in FORTRAN, distributed in punched cards and operated on an International Business Machines

(IBM) mainframe. According to Strasser, their general strategy was outlined by Ledley in a draft of his book – not published until 1965 – and then developed in cooperation with Dayhoff (Dayhoff and Ledley, 1962; Dayhoff, 1964; Strasser, 2010, p. 638).[5]

These practices of computer-assisted sequence analysis had a profound impact on the field of evolutionary biology during the second half of the 1960s. Suárez and historian Marianne Sommer have shown how, in 1965, Emilé Zuckerkandl and Linus Pauling postulated that protein sequences were 'documents of evolutionary history'. By measuring the similarities and differences between the sequences of insulin and other proteins shared by different species, they claimed it would be possible to determine their formation and differentiation over evolutionary time. Zuckerkandl, Pauling and other researchers proclaimed that sequence determination and associated techniques for molecular analysis of proteins were more objective measurements of evolutionary distances than the traditional criteria of evolutionary biologists (Suárez-Díaz, 2007, 2009; Sommer, 2008; quote from Zuckerkandl and Pauling, 1965, p. 357).

Sequence analysis algorithms became the distinguishing feature of molecular evolution as an independent field, differentiating it from systematics, taxonomy and other branches of evolutionary biology (Suárez-Díaz, 2009, pp. 47ff.; see also Hagen, 1999, 2001 and Dietrich, 1998). Molecular evolutionists adopted sequence comparison as an essential tool in their investigations and designed programs which overcame the 'alignment' and assemblage of overlapping sequence fragments. These programs were based on string pattern recognition algorithms, but incorporated tools that statistically calculated the 'degree of homology' between sequences. Evolutionary distances could then be inferred from the compared sequences, and phylogenetic trees constructed to represent the formation and differentiation of proteins in different species (Suárez-Díaz and Anaya-Muñoz, 2008, pp. 453–7).[6]

Throughout the late 1960s and the 1970s, Walter Fitch, Saul Needleman and Christian Wunsch designed sequence comparison programs with reduced memory requirements and adapted to the then emerging minicomputers. These apparatuses were of smaller size and cheaper price than mainframes and, despite frequent resistance, gradually colonised biological institutions.[7] With the spread of RNA sequence determination, the scope of comparisons was extended to nucleic acids. Graphics terminals were incorporated to the minicomputers, in order to correlate RNA sequences with their three-dimensional conformation (Suárez-Díaz, 2010, pp. 77ff.).

3.4   Biological computing and Cambridge molecular biology

Among molecular biologists in Cambridge, the use of these string manipulation and pattern recognition tools was rather limited. Kendrew's group included a significant number of staff in the Cavendish Laboratory – mainly women – in charge of the collection and organisation of X-ray diffraction patterns prior to computer processing. These data management practices became increasingly integrated into programs, but the main task of computers at the LMB Structural Studies Division – that is, the main goal that researchers pursued in using them – continued to be crystallographic calculations. During the late 1960s and 1970s, despite LMB crystallographers having their own in-house computer, calculations with large memory requirements were still performed in mainframes operated with punched cards or paper tapes, namely the Cambridge EDSAC-2 and an IBM at Imperial College, London (de Chadarevian, 2002, pp. 118ff.).

Kendrew attempted to introduce his group to protein sequence determination shortly before moving to the LMB, in order to link the amino acid sequence of myoglobin with its three-dimensional conformation. He hired Allen Edmundson, who moved to Cambridge in 1960, having completed his PhD at William Stein and Stanford Moore's laboratory at the Rockefeller Institute, where amino acid analysis had been automated (see Chapter 1). However, two years after his arrival Edmundson was transferred to Brenner and Crick's division, following the foundation of the LMB. Sanger declined to incorporate him, allegedly due to lack of laboratory space (id., 1996, pp. 370–3 and note 47).

Minicomputers and graphics terminals shared by small groups of researchers were introduced to Crick and Brenner's division during the early 1970s, mainly to process images of the nervous system of *C. elegans* taken with the electron microscope. John White, a PhD student under Brenner, devoted his thesis to the determination of synaptic transmissions between the worm's neurons and composed programs to reconstruct processes from series of pictures (White, 1974; White et al., 1976; García-Sancho, 2008, 2012).[8] However, the way in which Brenner framed the *C. elegans* project – within the idea of a genetic program – did not leave much room for computerised string processing. It was not until the late 1980s that the sequence data of the nematode was systematically collected (see Chapter 5).

Sanger did not introduce a specifically evolutionary approach into the protein work of his division. The two main lines of protein research in his laboratory during the 1960s were immunology, in the charge of

César Milstein, and identification of active centres of proteins, led by Brian Hartley. These investigations required comparison of sequences, but unlike in molecular evolution, this practice was restricted to small portions of the proteins where their functional activity – as antibodies or enzymes – was believed to lay. Milstein and Hartley developed techniques to detect, among the protein fragments, the locations of the disulphide bridges that linked protein chains and were involved in functionally relevant processes. The sequences of those fragments were then determined and compared manually (Frangione and Milstein, 1968; Hartley, 1970a,b).[9]

In 1968, two LMB researchers presented a survey on the use of computers within molecular biology. They made a distinction between the contribution of computing in determining the physical or the chemical structure of proteins: the former relating to the atomic three-dimensional conformation of the protein, the latter including the amino acid composition and sequence. Only two paragraphs of the paper were devoted to the chemical structure of proteins and the authors described a series of programs that automated the interpretation of amino acid analysis results. These results were presented in charts with peak graphs representing the response of the different amino acids to chemical reagents. The programs were able to correlate the peaks with the amino acid which had reacted to each chemical and, therefore, to determine the composition of the protein. The resulting data about amino acid composition was used to reconstruct the three-dimensional structure of the protein. The determination of the sequence required 'manual analysis' of the peak charts and the recording of data on a punched card (Gossling and Mallett, 1968, pp. 1090–5, quote from p. 1090).

The only researcher who systematically used and developed sequence analysis algorithms at the LMB was Andrew McLachlan, who joined the Structural Studies Division in 1971. Rather than coming from physics – the dominant background at that division – he had begun his career in medical research. McLachlan focused on comparison of protein sequences, with the aim of linking them to their three-dimensional folding. These comparisons allowed him to draw evolutionary conclusions, such as the correlation between amino acid repetitions in protein sequences and the duplication of genes in higher organisms (e.g. McLachlan, 1976; McLachlan, Bloomer and Butler, 1980).

During the 1970s and early 1980s, McLachlan designed programs for protein and RNA sequence comparison and presented their algorithms as 'based on those developed' by Needleman, Fitch and other molecular evolutionists (McLachlan, 1971, quote p. 409, 1976; McLachlan, Bloomer and Butler, 1980, p. 207). Since the comparison

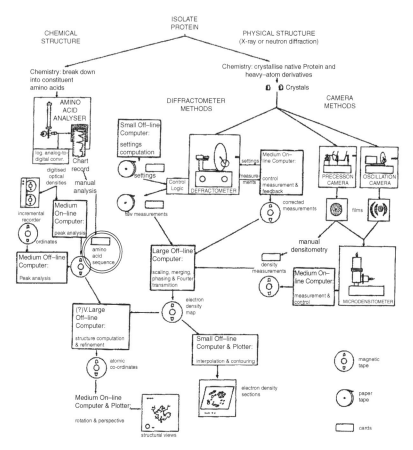

Figure 3.2   Use of the computer at the LMB during the late 1960s. The determination of the sequence of proteins (circled, towards the centre-left of the image) required manual analysis of the chart record produced by the amino acid analyser and the data was then recorded in punched cards. The rest of the computing practices pursued the reconstruction of the three-dimensional structure of proteins. (Gossling and Mallett, 1968, p. 1090. Reprinted with permission from T. Gossling.)

of sequences required relatively little memory, McLachlan ran his programs on a minicomputer, namely the IBM 360 (McLachlan, 1971, p. 413). This same minicomputer model had been introduced to the laboratories of Fitch and other early designers of sequence comparison algorithms during the previous decade (Suárez-Díaz and Anaya-Muñoz, 2008, p. 453).

Due to his isolation, the impact of McLachlan's work at the LMB was initially modest, leading Sanger's group to focus its search for computing expertise on other researchers. Barrell's first contact at the Structural Studies Division was Levitt, the crystallographer in charge of the latest programs for the calculation of physical structures. In the mid-1970s, three-dimensional mathematical modelling was perceived as the main computer application at the LMB, with sequence analysis being a secondary consideration. This was the reason for Levitt's rejection of Barrell's offer.[10] The alternative researcher Barrell approached after Levitt's refusal was also unconnected with the work of molecular evolutionists.

3.5   Sequencing software and the difficult formation of a community

Barrell's offer was finally taken up by Rodger Staden, a mathematical physicist 'interested in sequences' (Sanger and Dowding, 1996, p. 344). This interest, beyond three-dimensional structures, was to a large extent down to Staden's junior role at the Structural Studies Division in 1976. At the start of his career at the LMB – and lacking a PhD – Staden was in charge of writing FORTRAN programs for the repetitive processing of X-ray data.[11] Levitt, by contrast, had a more senior status and his own research projects, the specialised requirements of which were the aim of his computer programming (Levitt, 2001).

Staden designed his first programs for the assemblage and analysis of DNA sequences for Sanger's group between 1977 and 1980. As in McLachlan's case, they were not submitted to mainframes, but written and run with minicomputers. Minicomputers were becoming common as shared facilities in laboratories – including the LMB – but still coexisted with mainframes (Ceruzzi, 1998, chs 4–6). With the introduction of this technology, Staden intended that those doing the sequencing experiments would be able to 'run the software and take responsibility for their own data and its editing.' Technology writer, Glyn Moody, has argued that this introduction considerably helped overcome past resistances and familiarised members of Sanger's laboratory with the computer (Moody, 2004, p. 15).

Staden's first programs were run on a PDP-11, a popular minicomputer at that time which was commercialised by the US company Digital Equipment Corporation. As his previous crystallographic work had been, the programs were written in FORTRAN. Staden recalls that it was 'awkward to get the language to deal with arrays of characters, which it was not designed for'.[12] The operations these new programs allowed were similar to McCallum and Smith's previous computer aid in the sequencing of ØX-174. However, in Staden's programs the sequences – defined as 'character strings' – could be directly entered, assembled and edited by the researchers who had determined them (Staden, 1977, p. 4038).

These researchers would input the 'gel reading data' – the sequences they had read off the autoradiographs – into the minicomputer, which consisted of a keyboard attached to a large processor and was installed into a specific room shared by other LMB divisions. The partial sequences were printed on a continuous string of paper and the programs, when the relevant command was selected, searched for overlaps. Both the gradually assembled sequence and the remaining fragments could be saved on magnetic disks as files measured in kilobytes (id., 1979, p. 2601; see also id., 1978).

The files could be retrieved and edited. One program allowed the user to interactively search for any fragment by inputting its corresponding sequence. Staden's later programs translated the DNA sequence into protein and searched for codons – sequences that marked the beginning and end of genes, defined as protein coding regions in the DNA. In 1980, Staden coined the term *contig* and defined it as 'a set of gel readings that are related to one another by overlap'. The aim of his programs was to form a 'consensus sequence' whose length was 'the length of the contig' or, in other words, to reassemble the overlapping fragments into one single sequence (ibid.; quote from id., 1980, p. 3676).

Staden has admitted that, in those early years, he was unaware of the work of 'computer scientists' and consequently had to develop his programs 'from scratch'.[13] This retrospective statement appears to be true, since the only references in his single-authored papers up to 1979 were the articles in which Sanger's group presented the plus and minus, and dideoxy sequencing techniques (Staden, 1977, p. 4051; 1979, p. 2610). In 1980, Staden authored a new paper with an updated version of his programs and there he discussed the work that a group at Cold Spring Harbor Laboratory in New York had published the year before (id., 1980, p. 3694; Gingeras et al., 1979). Nevertheless, the 1960s and 1970s programs for protein sequence determination and comparison remained unacknowledged in all Staden's publications.

The group referred to in Staden's 1980 paper was headed by Thomas Gingeras and Richard Roberts. The latter had begun a career in molecular biology at Harvard during the late 1960s, initially focusing on RNA sequence determination. Roberts spent one month in Sanger's

```
SELECT OPTION BY NUMBER
STOP=0,ENTER=1,PRINT=2,DISPLAY=3,JOIN=4,COMPLEMENT=5,EDIT=6
  OPTION NUMBER = 4

LEFT CONTIG
  NUMBER OF LEFT GEL THIS CONTIG =232

RIGHT CONTIG
  NUMBER OF LEFT GEL THIS CONTIG =5

JOIN

RELATIVE POSITION IN LEFT CONTIG OF LEFT CHAR OF RIGHT CONTIG =596

          605.      615.      625.      635.      645.      655.      665.      675.
282  GAGTCAGGCGTCATTTTT1TGGTACGGAAAGTGATGCGAAAABCAGCGGCCAGTTVCGTVGAACAG
     GAGTCAGGCGTCATTTTTCTGGTACGGAAAGTGATGCGAAAAACAGCGGCCAGTTTCGTTGAACAG
     GAGTCAG4CGT1ATTTTTCTGGTACGGAAAGTGATGCGAARAAAACAGCGGCAGTCGTTGAACAGTCGCTGAGCCGACAGGCT6
       #  #                                # # ## ###### ### ## ####
       10.      20.       30.       40.       50.       60.       70.       80.
SELECT OPTION BY NUMBER
MOVE JOIN=1,EDIT LEFT CONTIG=2,EDIT RIGHT CONTIG=3,DISPLAY=4,COMPLETE JOIN=5,GIVE UP=6

  OPTION NUMBER = 2

INSERT=1,DELETE=2,CHANGE=3,RETURN=4
  OPTION NUMBER = 1

  POSITION =  639

  NUMBER OF CHARS =2

  CHARS TO INSERT INTO GEL  282 =___

INSERT=1,DELETE=2,CHANGE=3,RETURN=4
  OPTION NUMBER = 4

SELECT OPTION BY NUMBER
MOVE JOIN=1,EDIT LEFT CONTIG=2,EDIT RIGHT CONTIG=3,DISPLAY=4,COMPLETE JOIN=5,GIVE UP=6
```

Figure 3.3   A menu of Staden's early sequencing programs, which enabled the automatic detection of *contigs* – overlapping DNA fragments – as well as the editing and searching of sequence regions (Staden, 1980, p. 3688). The first versions were run in a PDP 11 (Programmed Data Processor) minicomputer, which lacked a screen and automatically printed the commands as the user input them. Reprinted with permission from *Nucleic Acids Research*. Copyright Oxford University Press.

laboratory learning the RNA techniques and in the late 1970s, having moved to New York, shifted to studying DNA, embarking on a sequencing project of Adenovirus-2. The computer programs described in Gingeras and Roberts's 1979 paper were framed by this project and personally endorsed by James Watson, co-discoverer of the double helix of DNA and then head of Cold Spring Harbor Laboratory (Roberts, 1993).

Roberts has recalled that in the late 1970s, connections between researchers engaged in computer DNA sequence analysis were weak, with early work appearing in a 'hodgepodge' of journals. However, *Nucleic Acids Research* gradually became the standard publication medium and published Staden's programs, as well as those designed at Cold Spring Harbor (id., 2000, p. 2). This journal was also in its early stages, being founded in 1974 as an 'interdisciplinary' platform for papers on the 'physical, chemical, biochemical, biological or medical properties of nucleic acids' (The Editors, 1974, p. III). Its editorial policy meant the generation of molecular biologists after Watson and the LMB founders dominated the early readership and contributors.

This new generation of researchers had begun their careers in the 1960s, during the golden age of molecular biology, and many of them had visited the LMB. Their early work was shaped by the emergence of recombinant techniques to alter the nucleotide sequence of DNA, and by Sanger and Gilbert's sequencing methods. These achievements – developed between 1973 and 1977 – led the younger breed of molecular biologists to increase their focus and reductionist perspective on DNA as the molecule that controlled all the processes characteristic of life. Watson, Crick and other self-defined founders of molecular biology blessed this new generation, as a means of extending their hegemony among the biomedical community, political committees and funding agencies. Researchers involved in recombinant DNA and sequencing achieved a growing scientific and socio-political authority, mainly within the emerging biotechnology market (see Chapter 6).

Roberts, Gingeras and other early developers of sequencing algorithms, thus, concentrated their efforts on nucleic acids and overlooked the previous work on proteins. In 1980, Gingeras and Roberts wrote a review on nucleic acid sequence analysis, which was published in a special issue of *Science* on the potentialities of recombinant DNA. The contributors, along with the journal editors, regarded the application of computers to sequence analysis as a consequence of the new techniques for the manipulation and determination of DNA sequences (Singer, 1980; Abelson, 1980). Therefore, Gingeras and Roberts only discussed nucleic acid programs that they knew 'either from publication or personal communication'. Their operations ranged from autoradiograph reading to gene search or RNA folding prediction. The sequence comparison and alignment algorithms developed by molecular evolutionists in the preceding decades were not referred to as earlier and relevant computer tools to handle biological sequences (Gingeras and Roberts, 1980, quote from p. 1327).

This review article represented a new case of appropriation of the history of sequencing by molecular biologists. As in the early 1960s Crick and Brenner had interpreted Sanger's move to the LMB as a conversion to molecular biology (see Chapter 2), twenty years later Gingeras and Roberts were omitting the protein side of computer-assisted sequence analysis. Gingeras and Roberts belonged to a second generation of molecular biologists who tended to see the history of sequencing in light of their disciplinary identity. This led them to consider sequencing and its computer analysis as techniques of an increasingly DNA-centred molecular biology, regarding the previous practices of molecular evolution and protein chemistry as rather alien to their interests.

Staden was initially exposed to this disciplinary view, but overcame it as he transformed his programming into a research problem. In 1980, he was appointed as a technician and formally moved to Sanger's division. Staden then decided to pursue an academic career and embarked on a PhD in computer-assisted sequence analysis. His thesis was jointly supervised by Sanger and McLachlan, who in the late 1970s had begun an increasing number of collaborations and mentorships of junior researchers involved in protein sequence determination.[14] McLachlan's supervision led Staden to appraise the previous algorithms of protein sequence comparison and connect them to his programs. It also expanded McLachlan's sequence comparison work to the field of DNA (e.g. Staden and McLachlan, 1982).

The Introduction to Staden's thesis included a 'historical perspective', in which he stated that 'proteins were the first sequences of biological molecules to be analysed using computers.' The thesis also reviewed the algorithms of, among others, Dayhoff, Fitch and McLachlan, and argued that with the emergence of DNA sequencing, this work had been 'rediscovered'. However, whereas in the protein programs the 'main interest was to compare the sequences', in DNA sequencing computers were used 'both during the elucidation of the sequences and in their subsequent analysis'. Staden intended his algorithms to become 'an integral part of the sequencing process' leading him to devote only one chapter of the 12 in his thesis to sequence comparison. The other chapters addressed sequence assemblage, different search strategies and analysis of certain DNA regions (Staden, 1984a, quotes from pp. 3–6).

In the course of Staden's doctoral study (1980–84), *Nucleic Acids Research* published two special issues on 'the applications of computers to research on nucleic acids' in which a number of authors used the word *software*. During the 1980s this term, along with *software package*, would come to replace *programs*. The papers in both these issues were dominated by DNA and RNA algorithms, but included programs that could be adapted for proteins. A number of researchers previously involved with protein programs had shifted to algorithms compatible with nucleic acids (e.g. Isono, 1982; Staden and McLachlan, 1982; Orcutt et al., 1982; Kanehisa et al., 1984). This suggests that McLachlan, and other researchers who had formerly designed protein programs, were struggling to overcome the narrow disciplinary approach to sequencing fostered by a DNA-centred breed of molecular biologists. Their strategy sought to form a community which included computer analysis of both protein and nucleic acid sequences through algorithms which, in many cases, were applicable to any sort of molecule.

Staden's work contributed to the formation of this community and in 1982, he presented an improved program for the 'automation' of sequence assembly.[15] His stated goal was to improve the efficiency of previous methods and avoid the task of comparing each newly sequenced DNA fragment against those corresponding to former sequencing work. He discussed protein alignment algorithms, among them a widely used one formulated by Needleman and Wunsch in 1970. When applied to the particular problem of DNA sequence assemblage, these previous algorithms became too 'general', since they required 'a lot of storage and computer time' (Staden, 1982, quotes pp. 4731 and 4742).

One of the alternative algorithms which Staden proposed estimated all the possible combinations between seven DNA nucleotides (16,348 in total) and placed them in a matrix in the form of single digits (combination number 1, number 2, number 3, up to number 16,348). The partially assembled DNA sequence and the newly sequenced fragments to match were then exhaustively analysed by the computer, which automatically assigned them their corresponding nucleotide combination numbers. If any of these numbers coincided, there was consequently an overlap, and all overlaps could be detected within a single computer analysis of the sequences. Staden presented the algorithm as follows:

> The sequence can be considered as a series of 7 character words that overlap each of their neighbours by 6 characters. Using an alphabet of 4 [the number of DNA nucleotides] it is possible to make 16,348 (4 to the power of 7) different words of length 7. If we have a table of 16,348 positions we can record whether or not each of the possible words of length 7 are present in the consensus [assembled sequence] .... In order to compare a gel reading [newly sequenced DNA fragment] with the consensus we simply look in the table to see if any of the gel's 7 character words are present. (ibid., p. 4743)

In this paper Staden quoted Jean-Pierre Dumas and Jacques Ninio, who had devised a similar algorithm earlier in 1982. Dumas and Ninio worked on DNA sequencing and RNA folding algorithms at the Department of Evolutionary Biochemistry, which by then was part of the Institute for

Research on Molecular Biology in Paris. They had presented a series of programs for 'performing pattern analysis tasks', in order 'to gain on computing time without sacrificing the quality of the results'. Dumas and Ninio acknowledged that they were adapting 'to the nucleic acid field some strategies' that were 'classical in lexicographic search problems', such as 'hash coding' and 'separate chaining' (Dumas and Ninio, 1982, p. 197).

These strategies enabled automatic searches for words starting with the same letters in a text, such as typing *pr** to search for both *print* and *process*. They had been used since the 1950s by programmers to operate with any string of characters, from programming instructions to bibliographies or biological sequences. Ledley had incorporated them in both bibliographic indices and protein sequence determination programs. String operation algorithms were of rather specialised use in computer science until the late 1970s and 1980s, when they became a regular feature of the then-emerging text processing programs, giving them wider visibility among the mounting number of computer users (Haigh, 2006a; Bergin, 2006a,b).

The increasingly efficient and diverse sequence analysis algorithms of the early 1980s demonstrate that the community engaged in computer support for DNA sequencing was gradually being penetrated, not only by previous protein work, but also by that of professional computer programmers. This development had required that McLachlan, Staden and other researchers approached sequencing and its computer analysis as research problems with a historical trajectory. The research problems that sequence computation embodied transcended – and had transcended in the past – a particular biological molecule or discipline. This perspective allowed the members of the community to overcome the previous design of both protein and DNA sequencing algorithms by largely non-interactive molecular evolutionists and molecular biologists.

Sequence assemblage in Staden's 1980s programs was infused with the practices of sequence alignment and comparison in molecular evolution, as well as by string manipulation in programming and text processing software. The programs reflected the transformation of the computer into an increasingly individualised device, engaged in the computation of strings of characters, rather than mathematically connected numbers. This use of the computer squared with the concept of genetic information that Sanger and Gilbert's techniques had fostered: a nucleotide sequence which could be determined by applying the new DNA sequencing methods. In Staden's programs, the practices of computer-assisted string manipulation were integrated into the form of work characteristic of sequencing, since the minicomputers in which the algorithms ran were operated by the same researchers who handled the techniques.

Specialisation in the computing dimension of sequencing provided Staden with an academic career and identity at the LMB. While at the Structural Studies Division he had merely reproduced an established process – computer handling of X-ray data; in Sanger's group he contributed to the formation of a growing community of practices.[16] This community, in charge of the improvement of computing tools, became increasingly distanced from the researchers who used these tools but focused more directly on the biological and chemical dimension of sequencing. The distancing and later reconciliation between these two communities was consolidated with the development of the first databases to store biological sequences.

4

Sequence Databases and the Emergence of 'Information Engineers'

Sanger's decision to introduce computers into his sequencing techniques coincided with a simultaneous effort to create a database for storage of the resulting DNA sequences. In the mid-1970s, when the first program to assemble and edit sequence data was being designed in his laboratory, Sanger had approached Olga Kennard, head of a widely reputed Cambridge-based repository of molecular structures derived from X-ray crystallography. He suggested to Kennard that they jointly establish a centralised DNA sequence database, one which would expand as sequencing spread. The project was discussed within the Laboratory of Molecular Biology of Cambridge (LMB, Sanger's home institution) but was never transformed into a concrete proposal. In the following decade, the LMB joined an international database initiative under the auspices of the European Molecular Biology Laboratory (EMBL).[1]

This chapter will investigate the early development of the EMBL database and its impact on sequencing as a form of work.[2] Unlike the emerging *sequencing software packages*, the database was designed and run by professionals external to academic biology and specifically hired for the task between 1980 and 1982. These professionals, who referred to themselves as *information engineers*, transformed the algorithms which had been designed for sequencing software into a series of tools able to capture the existing knowledge surrounding the DNA sequences stored in the database. The information engineers were instrumental in the consolidation of the community of practices engaged in computer sequence analysis, as well as the positioning of this community with regard to the EMBL molecular biologists. This positioning was problematic until the emergence of the field of biocomputing, institutionally incorporated into the EMBL during the mid and late 1980s to facilitate interactions between user biologists and the growing community of software and database designers.

4.1   The European workshop on sequencing and computers

The commitment of the LMB to the European database initiative was partly a consequence of the professional transfers and connections between this institution and the EMBL during the second half of the 1970s. The LMB researchers were especially active in the development of the European Molecular Biology Organisation (EMBO), which was founded in the mid-1960s and led efforts towards the creation of the EMBL in 1974. These efforts were marked by disagreements and difficulties in persuading European governments to become financially and politically involved. At the time of its creation, the EMBL was formed by 14 countries, all of which contributed proportionally to the funding of its operations. It was managed by a Scientific Advisory Committee of representatives of all the Member States and maintained strong connections with the EMBO, which ran a fellowship programme of research stays for European molecular biologists. Given that the Volkswagen Foundation had been a main funder of the EMBO during the mid to late 1960s, the EMBL was also receptive to private industry and engineering companies (Krige, 2002; Strasser, 2003).

The first Director-General of the EMBL was former LMB crystallographer John Kendrew, who led the planning of the Laboratory from 1971. Kendrew was involved in negotiating the international agreement that defined the intentions of the EMBL: (1) to pursue research on molecular biology that, due to the time-scale or investment required, was beyond the scope of national laboratories; (2) to develop instruments and technologies as centralised services for European molecular biologists, and (3) to provide teaching and training for researchers in the form of courses, workshops and visiting fellowships. As a result of this, the EMBL was divided into three divisions: Cell Biology, Structural Studies and Instrumentation. In its first Scientific Programme, produced in 1975, it was stated that 'the development of instrumentation' would be 'the largest division of the Laboratory's activities.' This powerful Instrumentation Division included, from the outset, a Computing Group into which Kendrew transferred the expertise he had acquired at Cambridge. The initial focus of this group, therefore, was assisting biologists in the mathematical calculations required for X-ray crystallography, neutron diffraction and other techniques.[3]

Shortly after his move to Heidelberg – where the General Headquarters of the EMBL were based – Kendrew persuaded Kenneth Murray to take a prolonged research stay at the Laboratory. During the mid to late 1960s, Murray had been a visiting fellow at Sanger's division and became the first researcher to attempt to determine DNA sequences at the LMB (see Chapter 2). Subsequently, he migrated to the Department of Molecular Biology at the University of Edinburgh – founded shortly after the LMB – and continued to use Sanger's techniques for his research on the hepatitis B virus (Hofschneider and Murray 2001). During his stay in Heidelberg, Kendrew offered Murray a position as Senior Scientist: he would be part of the committee in charge of defining EMBL policy and supervise younger visiting fellows.

Prior to Murray's arrival, he and Kendrew had actively cooperated in the organisation of recombinant DNA research at the EMBL. These techniques had begun to disseminate among molecular biologists shortly after the Laboratory's foundation and, during the second half of the 1970s, a specialised laboratory for recombinant DNA experiments – with strong security protocols – was built in Heidelberg. In 1978, when seven groups of EMBL molecular biologists were using recombinant DNA to isolate genes, Kendrew made clear his intention to make this 'one of the largest research groupings in the Laboratory.' These teams were instrumental in developing the then recently invented DNA sequencing methods of Sanger and Gilbert, which they used to characterise isolated genes.[4]

Murray's research stay at the EMBL began in 1979. One of his first initiatives was to organise a workshop on the use of computers in sequence determination, during which he hoped to expand computing expertise at the Laboratory to the handling of nucleotide sequences, as produced by the recombinant DNA groups. Murray initially invited to the workshop the leading European and US figures involved in DNA sequencing projects, including Sanger. Some of these researchers, however, forwarded Murray's invitation to their new collaborators in computing problems. The final list of participants included Rodger Staden, Thomas Gingeras and other representatives of the emerging community of DNA sequencing software creators, as well as some designers of the previous protein algorithms (see Chapter 3).

Murray's invitation letter emphasised the new opportunities that computer technology and programs offered in seeking 'correlations between sequences and biological features'. Among the topics to be discussed at the workshop, he referred to 'databanks and user centres', as well as 'the development and possible automation of methods for sequence determination'. The workshop agenda set 'sequence data banks' and 'sequence determination' methods as the main topics of the meeting. The attendees would also discuss the possible initiation of a large-scale sequencing project at the EMBL, such as the one Francis Crick had proposed to EMBO on the bacterium *E. coli.*[5]

well suited to this range of sequence reading. We are discussing the possibility of developing a totally automatic nucleotide sequenator, but before making a decision on this, we would certainly like to know what is being done elsewhere. We would welcome any information that you could pass on concerning individuals, or companies, you know with whom we might discuss this matter.

We are not yet prepared to commit ourselves to the determination of a major nucleotide sequence. There are attractions and temptations, but also considerable worries in what were termed projects K and H and at present we are not prepared to embark upon either. We do expect, however, to become increasingly heavily involved in DNA sequence determination and see little difficulty in providing useful material to feed any foreseeable work in the area of method development.

Yours sincerely,

Ken

K. Murray

Postfach 10.2209   Meyerhofstrasse 1
6900 Heidelberg   Germany
Telephone: (6221) 387 331   (direct)
            (6221) 3871 (via exchange)
Telex 461613 (embl d)

C. Automation. How far is this desirable and what are the limitations? Sample preparation, labelling and application to gels; fractionation and detection of oligonucleotides and comparative analyses of multiple gels.

7. The marriage of method development and automation to a problem of basic biological interest. Should the EMBL undertake the determination of the nucleotide sequence of the genome of *Escherichia coli*, or some other organism?

Figure 4.1   Above, Kenneth Murray's summary of the conclusions of the 1980 Schönau meeting. Below, extracts from the meeting's programme regarding automation of DNA sequencing and involvement in a large-scale sequencing project. (Graham Cameron's personal archive, European Bioinformatics Institute, Hinxton, Cambridgeshire, UK, folder on Schönau meetings. Reproduced with permission from K. Murray.)

Despite the EMBL's strengths in instrument development, the agenda for the workshop was ambivalent regarding the automation of sequencing. Beside the headline 'automation', Murray included the comment 'how far is this desirable and what are the limitations?' Both automation and the proposed large-scale project received little support during the meeting, held in April 1980 in Schönau, a small town close to Heidelberg. In a letter to the attendees following the end of the discussions, Murray explained that the EMBL would maintain its commitment to the development of sequencing methods, but while uncompromising about automation, he explicitly rejected the large-scale initiative (see Figure 4.1).

Looking at his post-Schönau letter retrospectively, Murray has stated that at the time of the workshop a large proportion of the attendees, as well as some researchers at the EMBL, were reluctant to embrace automation. When applying Sanger and Gilbert's techniques, these researchers wanted 'to decide themselves whether there came an A [adenine] or a T [thymine] in the sequence' rather than leaving the job to a machine. Regarding the large-scale project, Murray argues that sequencing and the recombinant DNA techniques had opened new research horizons among European molecular biologists, inside and outside the EMBL. It was, therefore, difficult to find biologists ready to 'put aside' their new research initiatives – derived from the interpretation and manipulation of DNA sequences – in order to embark on a project that merely aimed to accumulate data through the repetitive application of sequencing methods.[6]

The apprehension displayed about automation and large-scale sequencing at the Schönau workshop contrasted sharply with the more favourable welcome given to the database project. Key to this success was the form in which the EMBL presented the planned database: as a service to the European molecular biology community, rather than as a technology to be developed by the scientists themselves. During the workshop, a series of local protein and nucleic acid repositories created in the preceding decades – and similar to that which Sanger and Kennard had discussed in Cambridge – were considered. The attendees analysed the difficulties the researchers in charge of these collections were experiencing, especially in funding their maintenance and making the database work compatible with their investigations.[7] It was therefore decided that the future European database would be a centralised and international resource, maintained by specialised staff funded by the EMBL, rather than biologists foregoing other research projects.

4.2   Information engineers and the problem of interdisciplinarity

In his 1980, post-workshop letter, Murray stated that the EMBL had decided to develop a 'nucleotide sequence data library' under the sole condition that 'the necessary staff' could be recruited.[8] As the database had been defined as a service to the biological community in Schönau, Murray sought a specific kind of professional, far different from the EMBL molecular biologists. A job offer attached to the letter illustrated that the desired staff might not necessarily come from within established biological disciplines.

Candidates were required to have a background in 'mathematics, physics or computer science', the academic fields under which computing had been incorporated into biology during the preceding decades. They also needed to have used programming languages and carried out 'numerical and statistical analysis', but it was not compulsory for them to have a PhD. In fact, the selected applicant would be appointed as a Research Assistant or Manager, rather than the higher categories of Research Associate or Fellow, normally reserved for doctors. Biological expertise in 'problems surrounding nucleotide sequences and the structure of nucleic acids' was 'desirable', but not compulsory in the job offer.[9]

None of the professionals subsequently hired by the EMBL either held a PhD, or had extended biological expertise. The first database employee, Gregory Hamm, had studied a combined degree in biology and engineering and upon graduation, had worked in the emerging software industry of the United States. One of his previous roles had been in a small company, writing programs for various military projects. Hamm arrived in Heidelberg during the late 1970s on holiday. Having decided to extend his stay, he found a part-time job in sequencing software design within one of the recombinant DNA groups of the EMBL. By the end of 1980, he was awarded the database position, despite other candidates having PhDs in biology.[10]

Graham Cameron was hired to help Hamm in 1982, after the first release of the database. He had previously abandoned an undergraduate degree in Psychology and maintained a database of household information at the University of Essex (UK). Cameron was therefore equipped with considerable experience in database technology, which at that time was more a feature of industry and administration than academic biology. Cameron's expertise lay in compiling and organising large amounts of data, but he had no experience of biological or academic research.

Both Hamm and Cameron agree that biological skills were not essential during their early years at the EMBL. The latter defines himself as an 'information engineer' and claims the crucial issue in database technology was 'understanding information', it being a secondary consideration whether this information belonged to biology or any other realm.[11] For Hamm, the problems encountered when maintaining a database were those he had addressed in the software industry; a matter of engineering systems that did not require sophisticated biological expertise:

> I was probably the only one who was looking at [the database] as an engineering task rather than as a scientific task. Obviously, there was an important scientific content, it was necessary to understand the science, but basically I thought of it as an engineering task in which the problem was how to collect, edit, curate and distribute the body of scientific data around DNA sequencing. And I think this didn't require any new discovery about how nature worked; it required an awful lot of systematic work around handling and refining data.[12]

Hamm and Cameron belonged to an engineering tradition that had experienced considerable expansion from the 1950s onwards. Historian David Mindell has called it 'systems sciences' and defined it as an approach derived from different 'engineering cultures' which emerged in the late-nineteenth century and 'coalesced as [World War II] ended' (Mindell, 2002, p. 8). These cultures commonly conceptualised interactions between man and machine as a system marked by exchanges of information. With these information exchanges, the operator sought to lead the device towards a desired response. Military anti-aircraft batteries, early computers or the telephone were all, according to Mindell, inspired by the same principle: the operator entered a series of instructions and led the device to respond in a predicted way – with gunfire, the solution to an equation, or a call. Although each device developed independently prior to World War II, they all triggered multiple post-war applications – both inside and outside academia – under the common umbrella of systems sciences (ibid., pp. 7–11).

Hamm's defence software was one of these non-academic applications. His programs allowed the computer to recognise and connect the variables involved in the movement of a military target – such as shape of the target, speed and weather conditions – with the aim of predicting its behaviour. Missile or radar operators could therefore use the computer to determine future positions of their target, in order to either shoot or trace it. Hamm developed this system of data interconnection as a company employee rather than as a military expert. In 1980, after his EMBL appointment, he aimed to export this expertise to DNA sequences with no specific background in biology.

The application of systems sciences to biology was by no means new. Lily Kay's scholarship has shown that, following World War II, an increasing number of biologists had adopted different practices and models of systems sciences – particularly Claude Shannon's mathematical theory of communication – in order to study the problem of gene action.[13] These biologists aimed to predict the properties of genes mathematically, without the need to experimentally analyse them (Kay, 2000, chs 3–4; García-Sancho, 2007a, pp. 17ff.; Segal, 2003, part I and ch. 7; see also Quastler, 1953; Yockey, Platzman and Quasler, 1958). With the consolidation of molecular biology in the 1960s, researchers shifted to more experimental approaches but maintained the models of systems sciences within their conceptual framework. François Jacob and Jacques Monod, two leading molecular biologists, were inspired by the concepts of feedback and control – key to Norbert Wiener's cybernetics – in the design of their experiments to study the regulation of genes by enzymes (Creager and Gaudillière, 1996; Kay 2000, ch. 5; Rheinberger, 2006; Fox Keller, 1995).

The use of systems sciences in molecular biology was initially inspired by linear models of information transfer. Shannon's communication theory was designed to decipher encrypted messages during World War II, and presupposed a direct flow between the information source and the destination at the enemy front. After the war, the coding problem became a main line of research for molecular biologists and a perfect case for Shannon's one-directional information transfer from the source – the DNA molecule – to the destination – the protein. Shannon's model, additionally, squared with the dominant understanding of genetic information as a message or set of encoded instructions transmitted from the genes to the protein products (Kay, 2000, chs 4–6; Sarkar, 1996b).

However, as Evelyn Fox Keller has noted, the last third of the twentieth century was marked by technological transformations which affected the conceptualisation of biological processes: the telegraph was gradually substituted by the computer as the technology on which biologists modelled the functioning of the organism (Fox Keller, 1995). From the 1960s onwards, with the shift from the coding problem to more complex models of gene regulation, DNA expression began to be seen as dependent on multiple interrelated factors and, therefore, not fully captured by a one-directional transfer of information. This view led to the network structure of computer operation gradually emerging as a suitable alternative.[14]

Hamm's programming strategies were better adapted to this network perspective. The predictive power of his software was based on combining various sources of data rather than on a one-directional flow of information. When appointed as the EMBL database manager, he conceived this technology as a means to interconnect DNA sequences and order them according to other biological criteria. Unlike in Hamm's previous work in sequencing software, the database allowed accumulation of computer analyses of the sequences and used them to generate new knowledge. This led Hamm to present the EMBL database as a 'centrally supported' initiative to 'collect, organise and make freely available the published body of nucleic acid sequence data' (Hamm and Stüber, 1982, pp. 1–3). The database was no longer part of a locally applied sequencing technique, but the node of an emerging community of researchers increasingly focused on the determination and use of DNA sequences.

The transition from linear to network data processing models was favoured by a number of database applications of systems sciences from the late 1950s through to the 1970s. Following World War II, the increasing use of the computer in management led the owners of large civil companies to hire computer programmers and systems engineers to organise records spread among their multiple offices. Historian of technology Thomas Haigh has called these new professionals 'systems men' and defined their goal as the transformation of information management into a predictive science. They sought to create a 'totally integrated management information system', which would provide company owners with 'vital intelligence' about the firm (Haigh 2001, pp. 15–16). As in Hamm's programming, the systems men aimed to make predictions from the interconnection of different types of data.

Information management also penetrated public administration, becoming common in government offices, universities and libraries. The term *information engineer* – which Cameron used to describe himself – originated within this new expertise in business and civil service (ibid., p. 18; Kline 2006, p. 528). In his previous job in Essex, Cameron had become familiar with designing – or, in his own words, engineering – systems which centralised information and helped academic or government authorities to make decisions about the University's future. A database with entries about students, courses and grades could, for instance, be used to determine which courses needed to be dropped or redesigned, due to having the lowest number of students or the least satisfactory results.

Hamm and Cameron's entrance into the EMBL shows a further range of practices being introduced into sequencing as a form of work.[15] The origin of these practices qualifies a body of popular and academic literature which considers *interdisciplinarity* a distinctive feature of late-twentieth-century biomedicine. According to this literature, the emergence of new techniques to manipulate and determine DNA sequences fostered increasing interactions between molecular biology and computer sciences from the late 1970s onwards. As a result of these interactions, bioinformatics and other so-called new biosciences centred around DNA sequences emerged and became dominant disciplines with shaping impact on current biomedical research (e.g. Moody, 2004; Zweiger, 2001; Nakamura and Chow-White, 2011).

This scholarship overlooks non-disciplinary professionals who, like Hamm, Cameron or other systems men, were external not only to biology, but to the academic world as a whole. The practices they introduced into biological centres did not come from academic computer sciences, but from professional information management in the offices of private companies and public administration. Furthermore, these practices were not immediately combined with biology, since biological skills were unimportant in the original EMBL job offer and the initial design of the database.

Hamm and Cameron's non-disciplinarity reaffirms that the history of sequencing should be addressed from the perspective of practices, in order to transcend not only disciplinary boundaries, but also the artificial separation between the academic and non-academic worlds. The practices of data management they introduced cannot be fully explained from the perspectives of molecular biology, computer sciences and their integration into a field retrospectively entitled bioinformatics. This argument is further reinforced by the existence of biological databases that preceded Hamm and Cameron's effort, as well as the emergence of recombinant and DNA sequencing techniques. However, these prior databases had a more specific academic and biological origin.

4.3   Data management practices and earlier biological collections

Computer-based biological databases had existed since the mid-1960s, especially in fields which incorporated mainframe apparatuses. Margaret Dayhoff, the creator with Robert Ledley of the first protein sequence determination programs, and Olga Kennard, the researcher Sanger would approach for his proposed DNA sequence database, created repositories of protein sequences and structures derived from crystallographic computations in 1965.[16] These databases, according to Kennard, were founded on the 'belief that the collective use of data would lead to the discovery of new knowledge' which would overcome 'the results of individual experiments' (Kennard 1997, p. 159). Kennard and Dayhoff were, thus, introducing into biology the practices of computer-assisted interconnection of data derived from systems sciences.

These researchers, however, came from an academic background in biology and divided their time between database duties and work within their fields. The repositories they created were partially applied to their own research and partially circulated to other biologists. Kennard began her career at the Cavendish Laboratory during the late 1940s, shortly before Kendrew initiated his cooperation with mathematicians. She subsequently moved to the National Institute for Medical Research in London and was involved in setting up a database of X-ray structures at Birkbeck College, together with former Cavendish crystallographer, J.D. Bernal (see Chapter 2). This database was integrated by a number of cards with a row of linear round perforations at one edge. Various characteristics of the structures were assigned to different perforations, which were cut open, so that when knitting needles were pushed through various perforations and the stack lifted, the ones with the same features would fall out and could be grouped together. These collections of perforated cards – which were common classification systems at that time – inspired the punched cards later used in mainframe apparatuses.

In 1965, Kennard established the Cambridge Crystallographic Data Centre with the aim of developing the Cambridge Structural Database. This was a computerised central repository of organic and organometallic structures analysed by X-ray and neutron diffraction. Kennard never abandoned her career as a crystallographer and analysed the three-dimensional structures of various small molecules, including many of biological importance. She also used the information from the database to investigate topics such as the hydrogen bonds between the components of the stored molecular structures (Kennard, 1998; Strasser, 2011, pp. 67ff.).

Dayhoff's career has been investigated in detail by Strasser, who has shown how in 1965, as a result of her participation in the first computer programs for sequence determination, she began to publish the *Atlas of Protein Sequence and Structure*. The *Atlas* was a compilation of all the available protein sequences and appeared as a printed volume every one or two years. It included entries of sequences, along with papers in which Dayhoff and her team compared the data and constructed phylogenetic trees reflecting the evolutionary formation and differentiation of the proteins included in the volume (id., 2006, 2010, 2011, pp. 69ff.). This practice was common among molecular evolutionists involved in computer-assisted sequence comparison (see Chapter 3).

During the initial years of their databases, Dayhoff and Kennard used mainframes operated with punched cards. This created a distance between the compilation of the data, which was performed at their laboratories, and its retrieval and analysis at remote computer centres, some of them outside their home institutions. Even though minicomputers were spreading at that time, historian Joseph November has shown how their introduction into biological laboratories was especially problematic. The manufacturers attempted to foster it with the design of minicomputers such as LINC, oriented towards the specific necessities of biomedical researchers (November 2004, 2006, chs. 4–5). However, by the mid-1970s a significant proportion of laboratories remained non-computerised or, at best, delegated the work to external mainframes.

The foundation of the EMBL coincided with the emergence of more powerful, smaller and cheaper computers, such as the microcomputer and workstation (Ceruzzi, 1998, ch. 9; Campbell-Kelly and Aspray, 1996, Part 4). These devices – adapted to small groups of operators and reduced spaces, such as individual offices or laboratories – were quickly incorporated into the EMBL, renowned for its strengths in computing and instrumentation. At that time, however, biological researchers were not always keen on the complications of early in-house computers. This resulted in the further introduction of professionals such as Hamm and Cameron, with specific expertise in programming and handling the new technologies. Hamm and Cameron did not have to divide their time between research and computing work. The database was one of the first EMBL projects to be developed, from the outset, on microcomputers – which allowed the collection and analysis of DNA sequences to occur at the same location.[17]

Another key difference between Kennard and Dayhoff's efforts and the EMBL database, was their reception by fellow biologists and funding agencies. Kennard's collection emerged in a context marked by the Cold War and a fear that Europe was lagging behind the US and the USSR in the collection of scientific data. Her first grants were within a European initiative designed to counter that trend. However, with the gradual relaxation of international tensions she had to find self-funding strategies for her project to continue. During the 1970s, Kennard decided to lease the database entries to the pharmaceutical industry and large national research organisations, a scheme that ensured the continuity of her project to the present day. From the late-1980s onwards, the Cambridge Crystallographic Data Centre has been an independent charitable institution (García-Sancho, 2009, pp. 261ff.).

Dayhoff was initially funded by the National Institutes of Health (NIH), but given the insufficiency of the budget she had to seek additional support from, among other institutions, the US National Airspace Agency (NASA) and the Atomic Energy Commission. Like Kennard, she had to sell the *Atlas* volumes in order to guarantee the financial viability of the project. Strasser has shown that Dayhoff was never accepted as an equal by the community of academic biologists. Her efforts were seen as theoretical and framed in natural history, which at that time was considered irrelevant and oppositional to experimental biology.[18] Biologists were also unhappy that Dayhoff conducted research with the protein sequences they had determined experimentally and donated to the database for free. Exchanges of information within experimental biology, Strasser argues, were founded on the free circulation of data, once scientific publication had provided the authors with the necessary credit. Dayhoff's database challenged this 'moral economy', not only by being commercial, but by also making data available prior to publication (Strasser, 2010, 2006, pp. 114–18).

Hamm and Cameron's database emerged at a time when the collection and analysis of information was more valued by experimental biologists. The spread of increasingly in-house and individualised computers resulted in a phenomenon that social scientists have called the 'information society'. From the late 1970s onwards, governments and their citizens, together with businessmen and scientists, increasingly saw the control of, and access to, information – financial, techno-scientific or any other kind – as the primary means of increasing productivity, knowledge and social welfare (Castells 1996; Webster 1997; Harvey and McMeekin 2007; Parry, 2004; García-Sancho, 2007a,b; 2009). Historian of technology Ronald Kline has shown that these transformations contributed to the emergence of 'information technologies' as a category of socio-political debate. These information technologies were embodied by the personal computer, the initial models of which – Altair, Apple II and PC – started to penetrate, during the 1980s, into every social realm (Kline, 2006, pp. 520ff.; see Godet and Ruyssen, 1979; Danzin, 1979).

The growing scientific and socio-political concern with information gave Hamm and Cameron's database greater prospects than earlier efforts. The collection and analysis of DNA sequences using information technologies was perceived as an increasingly valuable practice by biomedical investigators and research funders. This perceived value was not only scientific, but also financial. The late 1970s and 1980s witnessed the emergence of the first biotechnology companies, supported via private capital and focused on the commercialisation of products derived from the analysis and manipulation of DNA sequences (Kenney 1986; see Chapter 6).

This scientific and socio-political enthusiasm fostered the EMBL initiative, which was followed by similar DNA databases in the US and Japan (Smith, 1990). The US endeavour, titled Genbank, took the form of a contract awarded by the NIH in 1982 to physicist Walter Goad and the company Bolt, Beranek, and Newman. Goad was based at the Los Alamos Laboratory, the institution in charge of the Manhattan Project to develop the atomic bomb during World War II (Lenoir and Hays, 2000). This Laboratory hosted a group of biologists and computer experts working on the effects of radiation on humans. Dayhoff also presented a bid for the NIH contract, which was unsuccessful. Her defeat had a profound emotional and professional impact on her, and she died prematurely, shortly after the first releases of the DNA sequence databases (Strasser, 2008, 2011).[19]

These databases differed from their antecedents by being part of top-down initiatives, and therefore benefited from relatively regular and sustained budgets. Their promoters made sequence submission agreements with journal editors which allowed the free and early circulation of data, making the databases compatible with publication credit requirements and other norms of the biological community (id., 2006, 2008, 2010, 2011). The more specific computer expertise of staff in charge of the projects affected how the databases were organised. In the case of the EMBL, the systems sciences background of Hamm and Cameron resulted in a series of algorithms, imported from other fields of computer programming, which adapted the standard database models to the specificities of the stored DNA sequences.

4.4   Sequencing algorithms, database technologies and operating systems

Hamm and Cameron's project, formally titled the Nucleotide Sequence Data Library, was preceded by important changes in database technologies. From the 1950s onwards, the database gradually abandoned its initial military connotations and became metaphorically associated with 'buckets', 'hubs' or 'pools' of data used by public administration and private offices. Throughout the 1960s and 1970s, the database acquired a more concrete identity, either as a collection of data to be fed into a mainframe or as a minicomputer file. Databases were gradually included in software packages with programs to automatically manage the entries (Haigh 2006b, pp. 33–4). Both these databases and programs were initially designed by computer manufacturers and then by the emerging software companies and divisions. International Business Machines Corporation (IBM) became one of the main early developers of database software (Campbell-Kelly and Aspray 1996, pp.174–6; Campbell-Kelly 2003).

Database management programs organised records according to various criteria. They established different models of interaction – hierarchical, network or relational – among the distinct types of stored data. If the data were, for instance, names of employees, payrolls and dates of birth, the programs allowed users to determine employees' ages and which of them received the highest salaries. The programs were adapted to the needs of the computing industry's main clients at that time – insurance companies, travel agencies, banks, libraries and job centres – all of which wanted a higher degree of control over the large volumes of data they produced (Campbell-Kelly 2003; CODASYL, 1969). During the 1970s, the relational model, which allowed unrestricted combinations between the entries, became established as the dominant data linkage criterion.[20]

None of these models was initially deemed suitable for the EMBL project. Hamm and Cameron, despite being aware of developments in database technology, did not enable interactions between the entries in the first releases of the Data Library, in 1982 and 1986. Their main concern at that time was making the database 'usable by human readers as well as by computer programs' (Hamm and Stüber, 1982; quote from Hamm and Cameron, 1986, p. 7). In this regard, they considered the standard available programs, which mediated between the computer analysis of the entries, their organisation and presentation to the user, to be unsuitable for the objects to be stored in the database, the DNA sequences.

The programs, according to Hamm, reflected a 'table view of the world', since they handled the database entries as self-contained data: for example, books borrowed by different library users with different return dates. The variable *books* could be related in many ways to the variables *user* or *return date*, but the text of the books could not be analysed. DNA sequences, by contrast, were different records, formed by continuous and large strings of already interrelated nucleotide units. The aim of the operator was not only to establish connections between the sequence entries, but also to find patterns within the strings in order to monitor their accuracy and attribute to them certain features. These practices were named 'polishing' and 'annotating', and became Hamm and Cameron's main endeavours in the early years of the EMBL database.[21]

During the first half of the 1980s, the EMBL team created a series of work routines on a day-to-day basis, which were then compiled into programs designed to systematise the management of stored sequences. Hamm initially interacted with a number of junior EMBL biologists associated with the database, and Kurt Stüber, a researcher at the University of Cologne and an enthusiastic collector of all kinds of objects. Stüber's main duties in Cologne were the design of RNA folding software and the maintenance of one of the local DNA sequence databases which had been reviewed in Schönau.[22] He was an unofficial part of the team until 1984, introducing Hamm to the adaptation of sequencing software for sequence database management. Hamm and Stüber redesigned the sequence comparison and assemblage algorithms which had became common during the 1960s and 1970s, and adapted them to annotate the stored DNA sequences and check for inconsistencies.

With the incorporation of Cameron in 1982, the database team members began to analyse literature on computer science more systematically, in search of improved and specific database algorithms. They also used a mid-1970s textbook, written by the Bell Laboratories programmers Brian Kernighan and P.J. Plauger, and employed by Hamm when designing sequencing software for the EMBL.[23] This book, titled *Software Tools,* described a series of algorithms Kernighan and Plauger used 'every working day' to automate a number of routine programming operations (Kernighan and Plauger, 1976, unnumbered page).

One of the main features of *Software Tools* was the adoption by the authors of the UNIX operating system. The Bell Laboratories had played a major role in the development of these systems, which emerged in the 1960s as mediators between programming instructions and their translation into computer language. The early operating systems incorporated text editors to assist programmers when they entered program instructions into time-sharing mainframes and later minicomputers. These instructions were translated into machine language and executed by the computer (Ceruzzi, 2005, ch. 3).

Routine	Author	Version	Date
Address Processing	A. Rudloff	1.1	11-OCT-1982
"	A. Rudloff	2.1	11-JAN-1985
Tape Distribution	A. Rudloff	1.1	12-OCT-1982
"	A. Rudloff	2.1	27-MAR-1984
Writing Data on Floppy Disks	G. Cameron	1.1	29-OCT-1982
Cover Note for Pre-release Data	GNC / AR	1.1	15-DEC-1982
Processing Pre-releases	GNC / AR	1.1	15-DEC-1982
Notes on Using CMS for Data Library Management	G. Hamm	1.1	29-JAN-1983
Author Review	KST / AR	1.1	23-JUN-1983
How to Deal with Programs and Data Submitted	GNC / GHH	1.1	5-JUL-1983
"	Pete McCaldon	1.2	23-OCT-1985
How to Change From One Release to Another	G. Hamm	1.1	24-JUL-1983
Program to Build File Names from GenBank Names	G. Hamm	1.1	26-AUG-1983
Notes on Assigning Accession Numbers	G. Hamm	1.1	17-OCT-1983
"	G. Hamm	1.2	5-APR-1985
How We Split the Journals	A. Rudloff	1.1	19-OCT-1983
How to Put Entries into the Library	G. Cameron	1.1	5-DEC-1983

Figure 4.2 List of work routines compiled daily by the EMBL team during the early development of the Nucleotide Sequence Data Library. A. Rudloff and Pete McCaldon were associated members of the database group. The initials *GNC* correspond to Graham Cameron and KST to Kurt Stüber. (Graham Cameron's personal archive, European Bioinformatics Institute, Hinxton, Cambridgeshire, UK, folder on memos and reports. Reproduced with permission from G. Hamm.)

UNIX was designed in 1969 and incorporated, as a novelty, a word processing program.[24] Kernighan and Plauger claimed they had both 'tested the programs' and 'typeset the manuscript' within UNIX (Kernighan and Plauger, 1976, p. 1). They described algorithms such as *find* or *print*, which could be used either in a text with programming instructions or a book written in a word processing program:

> Suppose you have to convert a 5000-line Fortran program from one computer to another and you need to find all the **format** statements.... How would you do it? One possibility is to get a listing and mark it up with a red pencil. But it doesn't take much imagination to see what's wrong with red pencilling a hundred pages of computer paper. It's mindless and boring busy-work with lots of opportunities for error. And even after you've found all the **format** statements, you still can't do much, because the red marks aren't machine readable.
>
> Another approach is to write a simple program to find **format** statements. This is an improvement..., [but] the trouble is that the program is so specialised that it will be used once by its author.... No one else will benefit from the effort..., and something very much like it will have to be reinvented for each new application.
>
> Finding **format** statements in Fortran programs is a special case of a general problem, finding patterns in text. Whoever wanted **format** statements today will want **read** and **write** statements tomorrow, and next week an entirely different pattern in some unrelated text. ... The way to cope with the general problem is to provide a general purpose pattern finder which will look for a specified pattern and print all the lines where it occurs. Then anyone can say **find** *pattern.* (ibid., p. 5)

Hamm, Cameron and Stüber, despite not using UNIX, incorporated a Virtual Operating System (VOS) into the design of their algorithms. The VOS granted the database the desired readability for both users and computers. In a 1984 memo, Cameron claimed that the ideal strategy for 'polishing' sequences was to combine 'general tools such as [text] editors and VOS filters' with specific programs to perform 'line checks'. He implemented 'pipes', that is, VOS commands which enabled the matching of different algorithms, and combined general purpose instructions with string pattern recognition tools, either designed at the EMBL, or previously used in sequencing programs. This combination allowed the database team to import professional programming practices of string matching, search and comparison into the Nucleotide Sequence Data Library.[25]

The EMBL database management algorithms were mainly used to automatically deduce features from the stored sequences and monitor their consistency. The features deduced – such as the location of the genes or proteins they coded for – were annotated in a particular section within each entry, entitled Feature Table. This meant the Sequence Data Library appeared different from the databases previously developed by the computer and software industries. It was no longer a table of self-contained interconnected data, but a more flexible structure whose entries were dominated by strings of nucleotides. From the automated analysis of those strings of nucleotides, the database could derive new knowledge which was added to other sections of the entries.

The materialisation of these string operation algorithms in the EMBL database represented a change of use with regard to previous sequencing software tools. The algorithms were no longer exclusively intended

```
ID    MMIG20    MUS.MUSCUL.IG.MOPC41; DNA; 350 BP.
XX
DT    82.01.01  (first entry)
XX
DE    First two exons in immunoglobulin light chain genes from
DE    cell line MOPC41.
XX
KW    differentiated gene; immunoglobulin.
XX
OS    Mus musculus (house mouse, souris domestique, Hausmaus)
OC    Eukaryota; Metazoa; Chordata; Vertebrata; Tetrapoda;
OC    Mammalia; Eutheria; Rodentia.
XX
RN    [1]    (bases 1-350)
RA    Altenburger W., Steinmetz M., Zachau H.G.;
RT    "Functional and non-functional joining in immunoglobulin light
RT    chain genes of a mouse myeloma";
RL    Nature 287:603-607(1980).
XX
FT    Key        From     To       Description
FT
FT    CDS        126      176      first exon (leader peptide)
FT    CDS        303      >350     second exon (variable part)
XX
SQ    Sequence   350 BP;  80 A; 82 C; 122 T; 66 G.
      CGTGACCAAT CCTAACTGCT TCTTAATAAT TTGCATACCC TCACTGCATC GCCTTGGGGA
      CTTCTTTATA TAACAGTCAA ACATATCCTG TGCCATTGTC ATTGCAGTCA GGACTCAGCA
      TGGACATGAG GGCTCCTGCA CAGATTTTTG GCTTCTTGTT GCTCTTGTTT CAAGGTTAAA
      ATGAAACTTA AAATTGGGAA TTTTCCACTG TTTCCAACTG TGGTTAGTGT TGACTGGCAT
      TTGGGGGATG TCCTCTTTTA TCATGCTTAT CTATGTGGAT ATTCATTATG TCTCCACTCC
      TAGGTACCAG ATGTGACATC CAGATGACCC AGTCTCCATC CTCCTTATCT
//
```

Figure 4.3 Sample entry of the second release of the Nucleotide Sequence Data Library (1986). The DNA sequence – at the bottom – is preceded by the Feature Table (FT), in which the location of two exons (genes) has been determined. (Hamm and Cameron, 1986, p. 8. Reprinted with permission from *Nucleic Acids Research*. Copyright Oxford University Press.)

to assemble or compare sequences, but also – and more importantly – to create a permanent record of what was known about them. The information derived from the pattern analysis of the sequences was collected in the Feature Tables and used to establish links with other biological knowledge, following the network approach to data adopted by systems scientists.[26]

This resulted in increasing exchanges and collaborations between the European database and other repositories. In 1987, the EMBL team agreed to unify formats and make the entries interchangeable with Genbank in the US, and the Japanese DNA sequence database. This was followed by an early 1990s agreement between the EMBL Data Library and SWISS-PROT, a Zurich-based repository which was becoming the largest protein sequence bank. Both databases interlinked their records via translation software which deduced protein sequences from the annotated genes in the Data Library entries (Bairoch 1991). The growing accumulation of knowledge around the database was also a consequence of the cooperation of Hamm, Cameron and Stüber with the EMBL biologists, overcoming the isolation they had suffered during the early years of their project.

4.5   Hierarchy, biocomputing and cooperation with biologists

The growing importance of information technology and DNA sequence analysis algorithms in the early 1980s did not completely erase the biases of experimental biologists towards database management. STS scholarship has shown how computer experts incorporated into biological centres at that time faced 'vast cultural differences' with the researchers already established in those laboratories (Cook-Deegan 1994, quote p. 285; Moody 2004; Chow-White and García-Sancho, 2012; Leonelli, 2010a,b). Molecular biologists based in the EMBL saw Hamm, Cameron and Stüber as 'secretariat', exclusively serving their research needs. As in the cases of Kennard and Dayhoff, their unawareness of the difficulties involved in database work led biologists to become impatient if sequences were not released immediately.[27]

This fracture between the biologists and the database team persisted during the initial stages of the Data Library and had a profound impact on sequencing as a form of work. The practices of gathering and ordering sequence data, performed by the emerging community of software designers and database professionals, became increasingly independent from their use. These professionals were perceived as hierarchically less qualified staff by the final users of the sequences, especially when, as in the EMBL, they lacked an academic background in biology.[28] For the EMBL biologists, the database team was providing a service in the form of data or raw material for research. The efforts to obtain such data were considered negligible or, at best, less important than the investigations to be conducted with them.

Hamm, Cameron and Stüber, thus, initially worked independently from the other EMBL groups. However, as the 1980s went on, discoveries such as alternative splicing, the complexities of gene regulation and the significant presence of introns – split genes – in DNA sequences of evolved organisms, made the simple day-to-day programs they designed increasingly obsolete. As the concept of the gene became more complex than that of a partial DNA sequence which coded for a protein, the algorithms inspired by text editors – the function of which was to annotate linear and uninterrupted nucleotide strings and translate them into protein sequences – became increasingly inappropriate.[29]

The database team attempted to refine its algorithms by incorporating staff with more advanced biological backgrounds. Internal reports show that between 1982 and 1985, biology students were employed and requests were made to the Laboratory's Director-General for the part-time involvement of senior EMBL biologists with the database. Database management was described in these requests as a 'production line' which included 'clerical staff' in charge of the literature searches for sequences, computer programmers responsible for algorithm design, and 'staff with biological knowledge' to complete the process. The main task attributed to the latter was the annotation and review of the sequences.[30]

The employment of biology students was insufficient and in the mid-1980s, the necessity of involving more senior biological staff became acute: the lack of sophisticated knowledge about DNA functioning was leading the database to experience a 'considerable backlog of data'. In 1985, Hamm wrote a circular to the EMBL biologists involved in sequencing with a direct and formal appeal for assistance. The tasks for which these researchers' help was requested were assessment of the interest of published sequences for inclusion in the database and the determination of which data should be included in the entries. Both clerical and student biological staff involved with the database either lacked this expertise or had a too 'generalist' background to perform such duties. The circular also raised the necessity of biologists acquiring 'some knowledge' of the conventions of the database, and proposed two sessions in which the student reviewer in charge of the annotations – Günter Stösser – would offer a briefing 'from the biologist's viewpoint'.[31]

Hamm's request was accepted by approximately 50 per cent of the researchers contacted. This led part of the EMBL biologists to better appreciate the intricacies of database work and gradually abandon their previous attitudes towards the Data Library team. It also coincided with debates between the Director-General of the Laboratory and the Senior Scientists of the different divisions in which the necessity of more systematised contact between biologists and computing staff at the EMBL – and perhaps even an organic integration of work – was raised.

From 1982, following Kendrew's retirement and the appointment of Lennart Philipson as the new Director-General of the EMBL, the Instrumentation Division had been divided into four units: (1) Physical Instrumentation; (2) Chemical and Biochemical Technologies; (3) Biological Instrumentation, and (4) Computing and Applied Mathematics. With this new structure, Philipson sought to 'complement' the existing EMBL strengths 'in physical instrumentation' with 'techniques and instruments stemming from developments in biochemistry, biology and chemistry'. The new Director-General justified the reform by stating that the 'most important technical advances in recent years' – namely the development of recombinant DNA – had occurred in the biological rather than the physical sciences. These achievements needed to be approached as 'extraordinary powerful and widely applicable' techniques rather than 'a biological problem area', as it had tended 'to be regarded at the EMBL'.[32]

As a result of this, the EMBL scientific programme for the years 1983 to 1987 established specific schemes for the development of techniques for recombinant DNA, sequencing and the production of monoclonal antibodies, in both the Units for Chemical and Biochemical Technologies, and Biological Instrumentation. The Data Library, previously based outside the Instrumentation Division, was relocated to the Unit for Computing and Applied Mathematics. In this new location, Hamm, Cameron and Stüber – like staff in the other Instrumentation Units – emphasised the development of techniques to be shared by the other Laboratory Divisions. The previous institutional setting of the database team, within the Structural Studies Division had, at times, led staff to present work as biological research projects rather than technical achievements. It was within the Instrumentation Division where Hamm, Cameron and Stüber refined their algorithms for the second release of the Data Library, in 1986.[33]

By that time, negotiations for the renewal of the EMBL scientific programme – which needed to be updated in 1987 – were already under way. In 1985, a workshop had been convened in Heidelberg, at which EMBL representatives and other European molecular biologists raised the necessity of a 'new category' of researchers well trained 'in biology and in information science'. It was decided to establish a 'fortified programme in biocomputing' at the EMBL, to develop methods of classification and data search, and 'redefine a matrix of biological knowledge'. In 1987, the Units of Biological Instrumentation and Computing and Applied Mathematics were merged into Biocomputing (European Molecular Biology Laboratory, 1987).[34]

This new Biocomputing Unit created an institutional framework which absorbed the practices and professionals configured during the preceding decades around the communities of sequencing software and databases.[35] Biocomputing as a field not only formalised these communities, but also enriched them through the incorporation of biologists at the former Biological Instrumentation Unit. This resulted in an increasing presence of professionals with a hybrid background, who could more effectively mediate between the expertise and identity of biologists and computing staff.

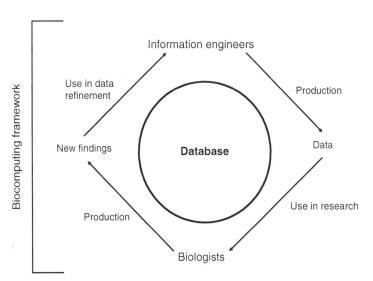

Figure 4.4   The database as a boundary object
Elaborated by author.

The emergence of biocomputing stabilised the EMBL database as a 'boundary object', with a multiplicity of actors and communities interacting around it (Leonelli, 2010b; Heeney and Smart, 2012; Star and Griesemer, 1989; Star, 2010). This interaction operated in a permanent and globally beneficial cycle, in which: (1) the database team produced sequence entries and these entries were then used by biologists in investigations which led to achievements, and (2) these achievements were the basis of subsequent refinements, of both the entries and the database algorithms, in light of collaborations between the biologists and biocomputing staff. The database, thus, embodied the latest state of knowledge about DNA sequences, knowledge which was permanently made available and updated through exchanges with the user biologists.

These entry management and sequence analysis algorithms also contributed to the mechanisation of certain practices of sequencing, namely the assemblage and annotation of DNA data. Such practices were increasingly controlled by sequencing software and databases within the new field of biocomputing. The way in which the user biologists faced their mechanisation – and the possible automation of other practices of sequencing – resulted in different materialisations of this form of work. As a result of one particular materialisation, the first automatic DNA sequencer emerged, becoming a commercial apparatus during the mid to late 1980s.

Part III

Mechanisation – 2: The Sequencer and the Automation of Sequence Construction (1980–2000)

> Molecular biology is sophisticated conceptually, but the routine tasks are repetitive, manual labor. The human genome project [HGP] is developing many automated techniques for molecular biology, including those for mapping and sequencing. In addition, thanks to the project, critical new computational tools for biology are being invented and the most sophisticated approaches in computer science are being applied to biology. (...) Moreover, the HGP is training a new type of interdisciplinary biologist who understands technology as well as biology. I believe these multidisciplinary biologists will be among the scientific leaders of the 21st century.
>
> Hood, 1990, p. 13.

The spread of sequence analysis software and databases did not end the essentially manual nature of sequencing as a form of work. Up to the mid-1980s, biologists still needed to perform the sequencing reactions in test tubes, carefully prepare gels, separate the radioactively-labelled DNA fragments through electrophoresis, make the autoradiograph picture, interpret the pattern of radioactive bands and, only then, use a computer. Sequencing software and databases were incorporated at the end of the process to input the sequence data, place it within the gradually assembled sequence, annotate and store the information.

Computing technologies were used only after the autoradiograph had been read-off and the researcher had deduced the sequence by scanning by eye the pattern of bands depicted on its surface (see Figure 2.3b).

During the first half of the 1980s, initiatives to also mechanise the interpretation of the autoradiograph emerged in the main sequencing technology centres, including the Laboratory of Molecular Biology of Cambridge (LMB, in the UK) and the European Molecular Biology Laboratory (EMBL, in Heidelberg). At the LMB, Rodger Staden, the designer of the first DNA sequencing software, devised a series of interfaces which automatically determined the nucleotide which each band represented, according to their position in the autoradiograph (Staden, 1984b). At about the same time, a group of the EMBL Instrumentation Division designed and commercialised an apparatus which could directly determine the sequence from the pattern formed by labelled DNA fragments (Ansorge et al., 1986, 1987).

Both these initiatives aimed to automate the reading off of sequences from a two-dimensional band pattern. The LMB and the EMBL teams based their efforts on a pattern of bands spread across the four lanes of a gel, which was the standard outcome of either Frederick Sanger or Walter Gilbert's DNA sequencing techniques. On the other side of the Atlantic, however, a simultaneous automation effort – one based on an alternative approach to sequencing – was being attempted. A team at the California Institute of Technology (Caltech, in Pasadena), led by Leroy Hood, had devised a 'sequenator' which automatically determined DNA sequences from a one-dimensional rather than two-dimensional band pattern. This device was based on Sanger's technique, but the DNA fragments obtained by applying the dideoxy reaction (see Figure 2.3a) were labelled with a fluorescent substance of a different colour, depending on the end nucleotide. This way, the fragments were differently labelled and could be distinguished after their separation in one gel lane rather than four. The sequence, thus, was automatically deduced from the linear succession of colours (Smith et al., 1986).

Part III of this book will argue that Hood's alternative approach derived from a different attitude towards the automation of sequencing when his group is compared with Sanger's. The Caltech team was not involved in the invention of sequencing and saw it as a repetitive laboratory practice, suitable to be automated. This resulted in a different role for computers in Caltech: instead of incorporating software and databases to handle the final data, as in Sanger's technique, the sequenator modelled the entire sequencing process on the workings of the microcomputer – introduced to Hood's laboratory at that time. Also, Hood saw sequencing and its automation as a business venture rather than a technique subordinated to the research necessities of the biologist – the dominant view within Sanger's group.

Despite the problems that comparative approaches present in historical research – and particularly in the history of science – their use is justified in this case study. A comparative approach between Caltech and the LMB will demonstrate that their differing uses of computing technologies resulted in a bifurcation of sequencing into two interrelated but independent forms of work. These different uses of computers were a consequence of divergent values and attitudes towards automation, partly shaped by the researchers' personas and partly by the history of their home institutions. As a result of this, sequencing diversified and became embodied in different instruments, which coexisted and circulated among biologists during the mid to late 1980s: an automatic apparatus commercialised by Applied Biosystems (ABI, a spin-off biotechnology company of Caltech), and manual protocols in which the LMB researchers explained to colleagues how to apply Sanger's techniques and their associated software.

The first half of Chapter 5 will develop this comparative approach and explain the roots of the divergent attitudes towards the automation of sequencing at the LMB and Caltech. The rest of the chapter will address more specifically the emergence of Caltech's one-lane approach to sequencing and the development of the prototype DNA sequenator. The transformation of this prototype into a commercial apparatus – the DNA sequencer – and its marketing by ABI will be the subject of Chapter 6. Both the design and the commercialisation of the sequencer, as I will show, triggered a shift in sequencing from a human-led to a mechanised form of work. Sequencing at Caltech and ABI involved public relations, marketing practices and customer support, rather than the biological and manual dexterity required in Sanger's laboratory. During the late 1980s – as ABI's first sequencer spread – it was not clear which of the two approaches would be implemented in a number of large-scale sequencing initiatives then proposed, among them the Human Genome Project.

My argument will be framed in the historiography of the connections between biological laboratories and companies. I will argue that the biotechnology market – particularly in ABI's case – represented a historically specific mode of association, one inspired by the previous information technology industry, which shaped the expectations and concerns raised by the first users of the sequencer (Swann, 1988; Quirke, 2007; Bud, 1993; Angell, 2005; Chow-White and García-Sancho,

2012). I will also draw on historical and STS literature on automation, factory models, the routine application of laboratory techniques and their impact on the governance and organisation of sequencing centres (Noble, 1984; Lynch et al., 2009; Hilgartner, 2004; Bartlett, 2008; Ramillon, 2007).

Among this scholarship, a paper by Peter Keating, Camille Limoges and Alberto Cambrosio on the reasons for the success of Hood's automation attempt will be particularly relevant (Keating et al., 1999). I will incorporate a historical perspective into their argument, building on a detailed reconstruction of Caltech's sequencing approach and its incorporation in ABI's apparatus. For this purpose, and as in previous chapters, I will reassess oral histories and retrospective accounts in light of the evidence collected from the archives and grey literature of Caltech and ABI. These materials will be compared with the records analysed in Chapters 1 to 4, and the papers and correspondence of Sanger's assistant, Alan Coulson, which are currently being catalogued at the Wellcome Trust Archives in London. Evidence from corporate literature, such as patent applications and the first instruction manual of the sequencer, will also be presented.

5
A New Approach to
Sequencing at Caltech

Attempts by Hood's group to automate DNA sequencing began in 1980 and produced a visible outcome in 1985, when the first prototype of the sequenator was unveiled. Their efforts contrasted with those of Sanger's team, where sequencing techniques were being applied to the DNA of different species – namely bacteriophage viruses – without attempts to expand automation. Hood and Sanger's agendas were profoundly shaped by their research ambitions and the histories of their home institutions. These personal and institutional differences framed the practices that Hood's team introduced into sequencing, and differentiated his approach not only from Sanger's, but also from other contemporary automation projects.

5.1 Contrasting groups, different personas

Hood led his own research group at Caltech from 1970, having previously spent three years as a senior investigator at the US National Institutes of Health. The focus of his team was immunology, the subject of Hood's PhD. During his doctoral years, Hood had been supervised by Caltech's biochemist William Dreyer, who was then involved in the development of protein sequence determination methods, along with the designers of the first computer programs for sequence assemblage, Robert Ledley and Margaret Dayhoff, who partially based their algorithms on Dreyer's techniques (Dayhoff, 1964, p. 112; Ledley, 1965, p. 379; see Chapter 3).[1] Dreyer's supervision resulted in a strong focus on protein sequence determination within Hood's group.

During the second half of the 1970s, when he was already established as a group leader, Hood collaborated with Dreyer and postdoctoral fellow Michael Hunkapiller to develop an automatic protein sequenator. Commercial sequence determination apparatuses existed from the beginning of that decade, the most popular being marketed by the chemical multinational Beckman Instruments.[2] Hood, Dreyer and Hunkapiller's device improved the speed of Beckman's sequenator and, crucially, allowed sequence determination with a hundred-fold less protein sample (Caltech, 1979, p. 43; id., 1980, pp. 40 and 52; Hewick et al., 1981).

This focus on instrument development within Hood's group coexisted with immunological research. A member of the team describes the laboratory at the beginning of the 1980s as 'two halves that were largely non-intersecting'. One half was a talented group of graduate students and post-doctoral researchers 'using molecular biology to study problems in antibody diversity'. The other group applied the protein sequenator 'to a wide variety of problems' and developed other instruments for the analysis of proteins and DNA. Both halves of the laboratory actively cooperated with 'groups all over the world'.[3] The collaborations involved engineers in the design of automatic apparatuses, among them those based in other divisions of Caltech.

Like many immunologists and biomedical researchers at that time (Podolsky and Tauber, 2000; García-Sancho, 2011a), Hood incorporated a genetic perspective to the study of antibody proteins after the generalisation of recombinant and DNA sequencing techniques. A research report of his group in 1975 considered the immune system a matter of 'organisation, expression and evolution of information in complex and multigenic systems' (Caltech, 1975, p. 76). As the 1970s advanced, the group members gradually established the cloning and sequencing of genes of immunological interest as their main goal. For this purpose, they adopted Sanger and Gilbert's techniques as soon as they became available, using them in combination with the developing protein sequenator (id., 1976–79).

In a series of public talks recorded during the 1980s, Hood announced a 'revolution in biology' based on the application of recombinant and sequencing technologies. Up to the mid-1970s, there was no way of directly accessing the 'information' stored in the DNA molecule, but the new molecular techniques would allow biologists to 'pick up the right word' and analyse its structure, in order to see how it dictated 'its instructions'. Thus, Hood defended the strong focus on instrument development within his group, since scientific achievements were always 'accompanied by advances in technologies'. He particularly described a 'magic quartet' of techniques being developed by his group: a protein and DNA sequenator, and a protein and DNA synthesiser.[4] Hood's group reported the first results on the automation of DNA sequencing in 1980 (id., 1980, p. 52).

The same public lectures allowed Hood to defend the involvement of biomedical investigators in the development of technologies as business ventures. Industry could provide valuable resources to academia, and biologists should take these opportunities – within ethical boundaries – 'as engineers and physicists had done in the past'. The important investments required to make instruments and technologies robust enough for the average biologist to use were what, according to Hood, made their production the natural province of companies and industry.[5]

These beliefs led Hood to seek private funding for his research and to organise large interdisciplinary teams. Throughout the 1980s, these included between 30 and 70 members and were highly competitive. One of the prime motivations for automating sequencing and synthesis in his laboratory was to accelerate the pace in which immune genes could be found and cloned, due to a strong competition with other groups. Dreyer, Hood, Hunkapiller and other collaborators filed patents for the protein sequenator and later for the synthesisers and DNA sequenator (Dreyer, 1976; Hood and Hunkapiller, 1979; Fung et al., 1985; Smith and Fung, 1986). In 1981, Hood created ABI to commercialise the automatic instruments developed at his laboratory.

Hood's personality and the orientation of his group stood in marked contrast with Sanger, who preferred small teams and maintained his identity as a biochemist within the LMB during the 1960s and 1970s (see Chapter 2). Sanger never intended to create or cooperate with a company, and his team freely distributed sequencing protocols and computer programs among other researchers.[6] During the early and mid-1980s, Sanger's group focused on applying the dideoxy method to various bacteriophage viruses and mitochondrial DNAs, including human (Sanger et al., 1978, 1982; Anderson et al., 1981). The only automation used in his laboratory was sequencing software, which had not been applied in Sanger's previous protein and RNA work. This software only mechanised the process of sequence assemblage and analysis: the practice of reading off the sequence was thus still performed by the researcher.

After Sanger's retirement in 1983, his assistant Alan Coulson moved to the neighbouring laboratory of Sydney Brenner. By then, Brenner's close associate, Francis Crick, had migrated to the Salk Institute in the US and the laboratory had been renamed LMB Cell Biology Division. Brenner's initiative to determine the role of genes in the development and behaviour of the nematode worm *C. elegans* (see Chapter 3) was still the main line of research for his group. However, Brenner's involvement in the project had gradually decreased, especially after his appointment as Director of the LMB in 1977 (Brenner, 2001, ch. 8).

A younger generation of members in Brenner's team progressively gained influence within the project. Among them was John Sulston, who had arrived at the laboratory in 1969 and conducted a detailed investigation of the embryonic and post-embryonic development of *C. elegans,* using a light microscope (Sulston, 1976; Sulston and Horvitz, 1977; Sulston et al., 1983; García-Sancho, 2008, pp. 102ff.). Coulson's move coincided with Sulston's decision to start a physical map of the worm's genome. The term *genome* had been used in Sanger's late 1970s papers to designate the whole DNA molecule in the cell nuclei of the viruses his group was sequencing (e.g. Sanger et al., 1978, p. 244; id., 1982, pp. 729, 734 and 735). *C. elegans* was the first genome of a multicellular organism to be addressed at the LMB, with the mapping effort formally begun in 1984 (de Chadarevian, 1998, 2004; Ankeny, 2001; García-Sancho, 2012).

Mapping genes of different organisms had been an aim of geneticists since the first decades of the twentieth century. T.H. Morgan and Arthur Sturtevant, who proposed the fly *Drosophila melanogaster* as a model organism, constructed linkage maps showing the location of genes within the fly's four pairs of chromosomes. They measured the degree of joint inheritance of certain features of *Drosophila* – such as eye colour or number of legs – and postulated that genes causing features which were jointly inherited were closer in the chromosome than those whose features were not passed together from parent to offspring fly (Morgan et al., 1925; Sturtevant, 1929). This strategy was used by Brenner to map the genes of *C. elegans* mutants during the initial stages of the worm project (Brenner, 1973, 1974), and by geneticists interested in disease inheritance patterns.

During the second half of the 1970s, geneticists and molecular biologists started using restriction enzymes and other recombinant DNA techniques to construct a different sort of map. Breaking various copies of DNA into overlapping fragments, they were able to assemble the fragments in their original order by detecting the overlaps. The result, called a physical map, allowed them to obtain short DNA fragments of a region of interest of a particular genome, which could then be easily sequenced or otherwise manipulated. Before Sulston's *C. elegans* initiative, partial physical maps of *Drosophila*, and of human genes causing mutations or diseases, had been produced (Lawn et al., 1978; Scott et al., 1983).[7]

Sulston's mapping effort was directed at the entire genome of *C. elegans*, rather than a particular mutation or region. Given the size of the worm's genome – 100 million nucleotide pairs – he decided to reorder, through a physical map, the thousands of fragments in which the template DNA would be split following its cleavage. To do so, he sought the cooperation of Coulson and adopted Sanger's sequencing instruments and strategic approach. The *C. elegans* mapping project coincided with similar initiatives by molecular biologists, on the large genomes of multicellular organisms such as yeast and the bacterium *E. coli*. All these projects began in the mid-1980s and continued through the following decade (Sulston and Ferry, 2002, pp. 43ff.; Olson et al., 1986; Kohara, Akiyama and Isono, 1987).

The *C. elegans* mapping method, known as the fingerprinting technique, began with the cutting of various samples of the DNA of the worm's genome into overlapping fragments. Each fragment was then inserted into a bacterium and submitted to molecular cloning: following division of the bacterium – and duplication of its genetic material – the inserted fragments also multiplied. The cloned fragments were cut again with different restriction enzymes which, as in Sanger's technique, always cleaved the DNA molecule at the same point in its sequence. Finally, the resulting sub-fragments – which had been radioactively labelled – were separated by gel electrophoresis and photographed in an autoradiograph. The outcome of Sulston and Coulson's mapping technique was, thus, the same as with Sanger's sequencing methods – an autoradiograph displaying a pattern of dark bands.

The practices that distinguished mapping and sequencing at the LMB were the preparation of the gel and the interpretation of the autoradiograph bands. Sulston and Coulson inserted the sub-fragments resulting from each fragment's cleavage into the different lanes of the gel, regardless of whether they ended in adenine, cytosine, guanine or thymine. Their aim was not to read off the sequence, but to compare the band patterns in each lane in order to find partial matches. If these were found, it meant the fragments overlapped: due to their sequence similarities, they had been cut at the same point by the restriction enzymes and the resulting sub-fragments had moved equally on different gel lanes. Sulston and Coulson sought to find overlaps in all the fragments and consequently to reconstruct their order and position from one side of the worm's genome to the other (Coulson et al., 1986; Sulston and Ferry, 2002, pp. 43–4).

Another area of transfer between Sanger's laboratory and the *C. elegans* mapping project was the computer software employed. In 1984, Staden released an updated sequencing program which partially automated the interpretation of the autoradiograph with a stylus connected to a computer. By touching each radioactive band, the stylus determined its position on the autoradiograph and transmitted this information to the computer, which gradually deduced the corresponding nucleotide

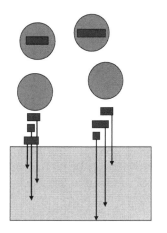

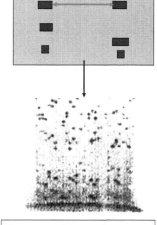

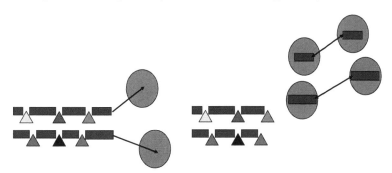

Various copies of the template DNA are cut into fragments. The fragments are then introduced into bacteria for cloning.

In dividing, the bacteria yield multiple copies of the DNA fragments.

The fragments are broken again and the resulting sub-fragments separated through gel electrophoresis (a technique in which the smaller move faster, and the bigger slower)

If common sub-fragments are detected on the gel, the fragments overlapped in the original DNA molecule (image from Sulston et al., 1988, p. 126. Reprinted with permission from *Bioinformatics*. Copyright Oxford University Press)

Figure 5.1   The physical map of *C. elegans*
Elaborated by author

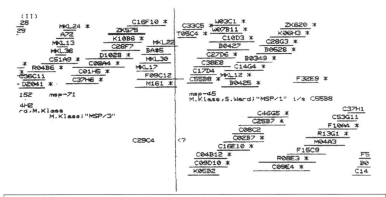

An ordered set of DNA fragments along the worm's genome – a physical map – is constructed from the overlap data. (Sulston et al, 1988, p. 128. Reprinted with permission from *Bioinformatics*. Copyright Oxford University Press)

Figure 5.1   Continued

sequence (Staden, 1984b). Sulston and Coulson actively cooperated with Staden, who also migrated to the Cell Biology Division after the mapping project had started. They adapted the software to detect common bands between the autoradiograph's lanes and, consequently, overlapping DNA fragments of the worm (Coulson et al., 1986; Sulston and Ferry, 2002, pp. 50–1).

In 1986, Sulston and Coulson presented the initial results of the mapping project. They defined the fingerprinting method and software as techniques 'for digital characterisation and comparison of DNA fragments', and described a 'data base' in which information about the fragments and their overlaps was being stored. As in Staden's sequencing programs, they referred to the overlaps between the fragments as 'contigs' (see Chapter 3; Coulson et al., 1986, quotes from pp. 7821 and 7823). The aim of the project, like that of the sequencing techniques, was to reduce the worm's genome to a single contig – a set of overlapping DNA fragments.

Another stated goal in Sulston and Coulson's paper was 'establishing genomic communication' between the growing international community of investigators engaged in *C. elegans* research (ibid., p. 7825). They did so by maintaining an intense correspondence with laboratories all over the world, from which Sulston and Coulson regularly received cultures of the worm's DNA fragments. Sulston and Coulson applied the fingerprinting technique and software, and by comparing the resulting autoradiographs with their database determined the position of the received fragments in the worm's genome. A printout of the corresponding contig was returned to the laboratory along with, if necessary, the mapping software and a copy of the database (García-Sancho, 2012).[8]

Sulston and Coulson, thus, freely circulated their mapping information and software to other researchers. In 1988, they were joined by Robert Waterston, a US researcher at Washington University in Saint Louis who had spent a sabbatical at the LMB while the mapping project was getting underway. Waterston shared the mapping effort and correspondence exchange with Coulson and Sulston, embracing their principle of free circulation of materials.[9]

Also in 1988, Sulston associated with other computing experts from the LMB crystallography division and developed an updated version of the mapping software. It incorporated a 'scanning densitometer', specifically built with the help of the LMB workshop (Sulston et al., 1988, 1989). This sort of scanning device had been introduced into sequencing software by Thomas Gingeras and Richard Roberts – the researchers with whom Staden had cooperated while at Sanger's laboratory – during the early 1980s (Gingeras et al., 1982, p. 104). It enabled the interpretation of autoradiographs without human intervention.

However, the LMB researchers considered that full automation of the *C. elegans* mapping effort was 'less powerful than the judgement of the operator'. They opted for a 'semiautomatic' approach in which the user had 'full knowledge of and control over the growing map'. The software displayed a number of proposed contigs on the screen, giving researchers the opportunity to revise the autoradiographs and validate the information (Sulston et al., 1988, quotes from pp. 125 and 132). This human control over data had been present in all sequencing software designed by Staden and in the previous programs for crystallographic calculations at the LMB (see Chapter 3). Gingeras and Roberts, who headed another major group in sequencing software, had also criticised programs which effectively separated 'the investigator from the examination' of his results (Gingeras and Roberts, 1980, p. 1324).

5.2   Institutional differences v national stereotypes

The contrast between the LMB researchers and Hood's group evokes historical stereotypes of science and technology in the UK and the United States. These clichéd images are founded on the belief that Britain gradually lost its industrial leadership because of the emergence of an anti-technological elite during the late nineteenth century. The historiography of twentieth-century science and technology in Britain is marked by debates about the *failure* of UK authorities to patent technologies such as antibiotic penicillin or monoclonal antibodies, which were subsequently manufactured and commercialised in the US (Wiener, 1981; Barnett, 1986; Mowery, 1984).[10] There is also a standard view of contemporary science and technology in the US, characterised by entrepreneurship and widespread contacts with industry. American scientists and inventors have long been portrayed – even in critical literature – as national myths or heroes (Noble, 1979; Hughes, 1989).

A closer analysis of Anglo-American contemporary science and technology shows not only that history is full of exceptions to these stereotypes,[11] but also that the Cavendish Laboratory – to a large extent the antecedent of the LMB – had a long tradition of entrepreneurship and contacts with industry. From the late nineteenth century onwards – when the Cavendish was predominantly a physics laboratory (see Chapter 2) – its researchers accepted paid assignments and actively cooperated with, firstly, the telegraphic industry and, later, electrical and engineering companies, in the contexts of World Wars I and II. However, they always emphasised the distinction between the University of Cambridge – engaged in academic research – and the 'work of a factory' – repetitive and routine endeavours (quote from Schaffer, 1992, p. 25; see also Sanderson, 1972, pp. 341–8 and Allibone, 1984).

This separation was not as rigid at Caltech, which had been created as a technical school in 1920. Its founders, Robert A. Millikan, G.E. Hale and Arthur Noyes, had played an important role in the organisation of US research during World War I and were all strongly opposed to public funding. They sought full support for Caltech's activities from within industry and local entrepreneurs (Goodstein, 1991, ch. 5). This led not only to an intense cooperation with industry, but also to the introduction into Caltech's structure of 'corporate models of management', as well as a 'problem oriented approach to science'. The Biology Division at Caltech, founded in 1928, had the drosophilist T.H. Morgan as its first director. He was chosen not only for his mastery of genetics, but also for his 'effective scientific management', and his contacts with the business world and private foundations (Kay, 1993, pp. 12–14).

This company-based approach resisted the shift towards state planning and funding of research in the US during World War II (Guston and Keniston, 1994). Despite Caltech being one of the main receptors of federal money for wartime research,[12] it maintained the policy of seeking additional private funds and introducing industrial expertise and practice into its investigations. During the post-war and Cold War years, Caltech's activities were supported 'through a wide network of grants and contracts from government, pharmaceutical companies and agricultural industries, as well as by gifts and endowments from Southern California's business community' (Kay, 1993, pp. 236–7).

The diversity of funding sources at Caltech profoundly shaped the development of Hood's group. By 1970, when the team at the Biology Division was created, Caltech already had an extended collaborative network with charities, industry and public institutions. The annual reports of Hood's team show that between 1975 and 1987, its research activity was supported by a pool of federal agencies, the National Institutes of Health (NIH) and the National Science Foundation; by companies, Beckman Instruments and AmGen; and by private foundations, the American Cancer Society and the Arthritis Foundation (Caltech, 1975–87).

According to Hood, the project to automate DNA sequencing was exclusively funded by private money, mainly from his start-up company, ABI. By the time the prototype DNA sequenator was presented in 1985, Hood had become one of the main supporters of sequencing the human genome, a possibility beginning to be discussed at international biomedical meetings. In 1990, when the US Administration announced a concerted Human Genome Project (HGP), Hood referred to it not as *big science*, but as 'interdisciplinary science'. However, he considered big science projects – those involving large, multidisciplinary research teams, and attracting academic and industrial funding – as perfectly suitable and convenient for biology (quotes from Hood, 1990, p. 13; on big science see Lenoir and Hays, 2000; Beatty, 2000; Galison and Hevly, 1992).[13]

Hood left Caltech in 1992, following disagreements with its authorities regarding the size of his group and volume of research contracts. His next institutional setting was the University of Washington at Seattle, where he founded and chaired a large-scale Department of Molecular Biotechnology, supported by the prominent Seattle-born computing entrepreneur, Bill Gates.[14] There, Hood developed another automated instrument: an ink-jet synthesiser to produce DNA arrays (Hamilton, 1991; Hood, 2008, pp. 1.17ff.). The Department of Molecular Biotechnology split from the University of Washington in 2000, and was transformed into the Institute for Systems Biology, a private non-profit institution still operative and chaired by Hood today (2012).

The institutional orientations of Caltech and Hood's group differed markedly from those at Cambridge. The conception, planning and foundation of the LMB were the result of negotiations, during the mid- to late 1950s, between a growing group of biologically oriented researchers at the Cavendish Laboratory, and the Medical Research Council (MRC), one of the scientific bodies of the British Government (see Chapter 2). The MRC had been founded in 1913 in connection with new arrangements for national medical insurance, particularly in relation to tuberculosis, but the Council soon became dominated by academic scientists rather than clinicians, and in the 1920s strongly favoured basic biomedical research (Austoker and Bryder, 1989; Timmermann, 2008). Walter Fletcher, a Cambridge physiologist and the first MRC Secretary, played a major role in the foundation and early development of the Dunn Institute, later the Department of Biochemistry at Cambridge and Sanger's initial home institution until his move to the LMB (Kamminga and Weatherall, 1996; Weatherall and Kamminga, 1996; Kohler, 1978; see Chapter 1). Before World War II, most of the British biomedical establishment was further removed from industry than were British chemists or engineers, though there were links with new industrial pharmaceutical laboratories, notably those created by the Burroughs Wellcome company.

This situation persisted with the international boost to basic biomedicine which followed World War II. As a result of this, during the planning of the LMB in the mid to late 1950s, the MRC opted for a funding scheme based on trust and long-term investments. MRC officers made personal agreements with the heads of the future LMB divisions – all awarded the Nobel Prize between 1958 and 1962 – and funded their research over extended periods of time. This was achieved through block grants which were monitored in periodical visits and reviews. In 1963, one year after the foundation of the Laboratory, the MRC requested that Sanger and the other group leaders, Francis Crick and Max Perutz, predict their financial and spatial necessities up to their retirement, 15 years in the future. This resulted in a proposal for an extension of the LMB, in which Brenner first presented his *C. elegans* project. The document described the different lines of research at the Laboratory and future plans in a rather informal style, not supplying complex details. This language contrasts with present-day standards in the writing of applications.[15]

The MRC funding scheme materialised in a particular *culture* or environment which permeated both the LMB members and their scientific efforts. This contrasted with Caltech, which organised its activity around concrete and applicable research projects, funded by contracts with charities, companies or the US Administration. Hood and other Caltech researchers – some of them also Nobel Prize holders – needed to assume an entrepreneurial role and seek private or public support for their work, which had to be oriented towards tangible outcomes.

The LMB members, by contrast, had more financial security and freedom to decide – within each block grant period – the orientation of their research, without worrying about its applicability. Sulston has suggested a 'tradition' at the LMB based on long-term academic work, 'not done in pursuit of any ulterior motive, financial or otherwise' (Sulston and Ferry, 2002, p. 14).[16]

The LMB tradition squared with the generous public expenditure in science and social policies in Britain during the 1960s. In the following decade, however, the budget for scientific research was drastically reduced and the Government forced its councils to sign goal-oriented contracts with the institutions they decided to fund. Both the LMB and the MRC fiercely challenged this decision. MRC officers sought to escape from the new policy and in 1981 managed to return to a long-term and non-project-oriented funding scheme. They were the only British council to do so, in spite of the strong neo-liberal ethos of Margaret Thatcher's Government, elected two years earlier (de Chadarevian, 2002, pp. 339–53; Wilkie, 1991, pp. 81–8).

This commitment to basic and publicly-oriented science became a main leitmotiv of the LMB Faculty members, despite significant personal differences. Sanger's tenure at the sequencing division was characterised by the development and free circulation of new methods, emulating the traditional experimental strategies of molecular biology (see Chapter 2). His retirement in 1983 was largely prompted by the lack of scientific satisfaction he gained from the repetitive application of DNA sequencing techniques, and his resistance to developing an automatic apparatus (Sanger, 1988). Sulston, by contrast, spent a significant part of his career patiently applying the fingerprinting method – inspired by Sanger's techniques – to the construction of a physical map of *C. elegans*. However, as this mapping effort increased in scale, he moved to a new institutional environment.

Sulston was appointed Director of the Sanger Centre in 1993 and left the LMB – and Coulson – for the new post. The Sanger Centre was built to host the British participation in the HGP on a new research campus in Cambridgeshire. It also hosted the Biocomputing Unit of the EMBL – renamed the European Bioinformatics Institute – and a recently initiated project to sequence the DNA fragments of the still partial map of *C. elegans* (see Chapter 4; Sulston et al., 1992). The HGP had been announced in 1990 as a concerted effort, led by the US, that would integrate a number of national initiatives. In Britain, it was mainly funded by a private charity, the Wellcome Trust, with the MRC only contributing part of the budget for the new Sanger Centre (Fletcher and Porter, 1997).

Sulston, however, became one of the main defenders of the public nature and universal availability of the resulting human genome sequence. Shortly before the foundation of the Sanger Centre, he and Waterston – Sulston's *C. elegans* counterpart in the States – had been approached by Hood and US entrepreneur, Frederick Bourke. They proposed that Waterston and Sulston should head a company in Seattle which would finish the sequencing of *C. elegans* and start on the human genome. After some deliberation, Sulston and Waterston jointly rejected the offer: not only would it relegate *C. elegans* in favour of the more practically-oriented human sequencing, but more importantly, the DNA sequences would be patented. In his memoirs, Sulston retrospectively argues that he believed he would simply have been the scientific director 'of a commercial sequencing organisation' and did not see 'the point of making money for its own sake' (Sulston and Ferry, 2002, pp. 101–2).

Sulston and Waterston's decision was endorsed by a number of researchers in both Europe and the US. Among their main supporters was James Watson, co-discoverer of the double helix of DNA and considered to be one of the founders of molecular biology. In 1989, Watson had been appointed Director of the National Human Genome Research Institute, created by the NIH to host the US human genome sequencing initiative. He had devoted part of the NIH human genome budget to support the initial stages of the sequencing of *C. elegans,* and did not want to lose the project. When Hood and Bourke's offer arose, he persuaded British institutions to keep the worm's sequencing results within the public domain. Finally, the MRC and the Wellcome Trust committed to the funding of both worm and human sequencing at the Sanger Centre (García-Sancho, 2012; Sulston and Ferry, 2002, ch. 3).

5.3   Research values and attitudes towards automation

The different trajectories of Sulston, Sanger and Hood suggest that the research priorities of their groups were the result of divergent values attached to the practices of science and technology, partly shaped by the institutional histories of Caltech and the LMB, and partly by their contrasting personas. For LMB Faculty members, science always implied personal involvement, and therefore technology was to be subordinated to human operators. By contrast, the Caltech group believed science could and should be fully automated; they consequently envisaged technology as both a scientific and commercial opportunity, in the form of an autonomous apparatus which could be used by researchers and marketed by a company. There were individual differences within the LMB and Caltech researchers, which were to some extent generational. Sanger and Sulston held different attitudes towards large-scale projects, but unlike Hood, they shared a belief in the public nature and universal availability of science.

We have seen that Sanger, Sulston and Hood's values were reflected in divergent attitudes towards automation and embodied in different mapping and sequencing instruments: on the one hand, the LMB protocols and software always remained under the user's control; on the other, Hood's machine eliminated human intervention in a large part of the sequencing process. Caltech's prototype sequenator was protected by patents, to transform it into a commodity to be marketed by ABI. Conversely, the LMB protocols and software were freely distributed to other researchers.

The contrast of values and attitudes between the LMB and Caltech does not simply reflect stereotypical differences between Europe and the US. The perspective of the LMB was generally shared by researchers based at institutions with a strong tradition in academic biology, even in the US – such as Watson, Gingeras and Roberts at Cold Spring Harbor – or laboratories proudly following the purely basic research approach defended by the *founders* or *heroes* of molecular biology – for example Waterston's group at Washington University in St Louis. Conversely, Caltech's approach to sequencing was adopted by new institutions with a strong focus on engineering and instrumentation, including European examples such as the EMBL, which also developed a fully automated DNA sequenator (see below).

The protocols and the sequenator should thus be seen as the technologies in which a particular institutional and personal style of conducting research was historically materialised.[17] This historical dimension suggests that manual and automatic sequencing developed as progressively differentiating rather than mutually exclusive forms of work. Members of Hood's group used manual sequencing protocols extensively before embarking in the DNA sequenator, and their first attempts at the automation of sequencing were based on Sanger and Gilbert's techniques. However, given that these techniques had been designed to be applied by humans, the Caltech researchers gradually built an alternative sequencing strategy, one more suited to full automation. A reconstruction of how this strategy emerged will enable me to show how the contrasting styles of research at the LMB and Hood's group resulted in different incarnations of sequencing as a form of work.

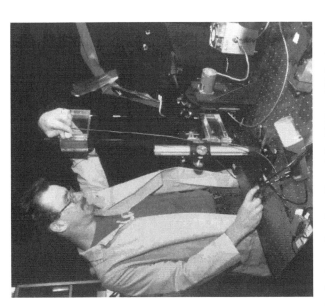

```
CHAIN TERMINATOR SEQUENCING
---------------------------

1.  Introduction

2.  Materials

3.  Stock solutions and buffers:  A for gels

B for reactions

4.  Preparation prior to sequencing

5.  Gel preparation

6.  Primers

7.  General techniques

8.  Sequencing procedure and notes

9.  Gel electrophoresis and autoradiography
```

Figure 5.2 The instruments in which the differing values and attitudes towards automation at the LMB and Caltech were respectively embodied: a sequencing protocol (left) and a fully automated sequenator (right). (Papers and Correspondence of A. Coulson, Wellcome Trust Archives, London, UK, provisional reference number PP/COU, and courtesy of Lloyd Smith. Reproduced with permission.)

5.4   Out of the 'autoradiograph world'

Manual DNA sequencing protocols were introduced into Hood's immunological investigations as soon as they became available, between 1975 and 1977. Researchers in his laboratory found them complementary to protein sequence determination in the study of both antibodies and the genes in charge of their synthesis (Caltech, 1977, p. 48; 1980, pp. 43–4). Unlike Sanger's group, which enjoyed the practice of sequencing and found a certain 'pleasure' in reading off autoradiographs (Sanger and Dowding, 1996, p. 344), the Caltech researchers considered the manual techniques monotonous and wished to employ the time devoted to sequencing in other aspects of immunology. The competitive atmosphere of the group led its members to see the automation of sequencing as an opportunity to both strengthen the ongoing line of research on instrumentation, and to isolate and clone immune genes faster than rival laboratories within the 'hunt' for genes then spreading through biomedical research (Kevles, 1985; Harper, 2008; Cook-Deegan, 1994, ch. 2).

The first attempt at automating DNA sequencing by the group was performed by Henry Huang, a biochemist with an interest in engineering who had joined Hood's laboratory in the late 1970s. Huang had previously been based under Dreyer, Hood's former PhD Supervisor, and had formulated the 'area-code hypothesis' in cooperation with those two researchers. This theory postulated that, during embryological development, immune cells possessed a series of membrane molecules which acted as 'distinct cell surface addresses' and were the basis for the 'cell-cell recognition' process which occurred in the conformation of the immune system (Caltech, 1977, p. 44; Hood, Huang and Dreyer, 1977). Cells with the same membrane molecules, though different, were able to recognise each other, as one might recognise an area code, despite the subsequent phone number being unknown. This hypothesis suggests a tradition of modelling biological processes on the information technologies used daily by Hood and Dreyer's group, in this case the telephone.

Huang's automation strategy, devised in 1980, was based on Gilbert's DNA sequencing technique and attempted to eliminate human intervention in the interpretation of the sequence. Instead of basing such interpretation on the radioactive bands of a picture of the gel – the autoradiograph – Huang sought to detect the corresponding DNA fragments on the gel itself after their separation. He exposed the gel to a source of ultraviolet radiation and measured the degree of light absorption by each DNA fragment. The measurements were meant to automate the interpretation of the sequence by a similar process to that of the manual reading off of the autoradiograph (Caltech, 1980, p. 52). However, the signals derived from the gel were too weak for a precise estimation and Huang left Caltech in 1982, without completely solving the problem. Upon his departure, Hood's group ceased using the technique (Cook-Deegan, 1994, p. 65).

The automation of DNA sequencing was soon taken up again by Lloyd Smith, and Michael and Timothy Hunkapiller. Michael Hunkapiller was one of the most senior members of the laboratory and led the part of the team devoted to instrument development. He had played an important role in the design of the protein sequenator and was considered to be Hood's right hand man. His younger brother, Timothy, was a graduate student who had recently joined the group; interested in the use of computers in the laboratory, he was devising algorithms for protein and DNA sequence analysis. Smith was beginning a postdoctoral stay at Caltech following a PhD in biophysics at Stanford University. He had been working on the measurement of the lateral diffusion of lipids and proteins in cell membranes using laser radiation (Hood and Hunkapiller, 1979).[18]

Smith and the Hunkapiller brothers decided it was necessary to devise a new approach to DNA sequencing rather than continuing to pursue the automation of gel or autoradiograph interpretation. The main reasons for this were the difficulties in detecting DNA fragments on gels and the inconvenience of the band pattern formed on autoradiographs. On autoradiograph pictures, the labelled DNA fragments appeared as dark radioactive bands distributed along four lanes (see Figure 2.3b). The Caltech researchers did not feel comfortable with the use of radioactivity[19] and considered this pattern especially complicated for automatic interpretation: some of the autoradiograph bands were blurred due to their imperfect separation and they often *smiled*.

The smiling effect meant the DNA fragments located on the central lanes of some gels moved faster than those at the edges. This made it difficult to extract sequence data from the autoradiograph, since the central bands were unjustifiably below their neighbours. Whereas the trained eye of a researcher could easily correct their positions and read-off the sequence, an automatic device had enormous difficulties accounting for such an unpredictable movement, one which did not occur systematically in all the gels.[20]

The scanning interfaces devised by Sulston, Gingeras and Roberts attempted to systematise this randomness and automatically process the autoradiograph according to predictive algorithms. The Caltech researchers believed a more promising approach would be to partly modify the sequencing procedure, in order to produce an alternative outcome. When reflecting retrospectively about this shift, T. Hunkapiller has claimed they were abandoning the 'autoradiograph world' and entering into a 'reappraisal of the whole sequencing process'.[21] With the term *world,* he indirectly acknowledges the whole repertory of practices, attitudes and values – both institutional and personal – which constituted sequencing as a form of work established by Sanger and Gilbert. In leaving the *autoradiograph world*, the Hunkapiller brothers and Smith were constructing an alternative form of work, more suited to their belief in non-human-mediated automation. They were replacing the chemical and biological dexterity required in Sanger's sequencing, and in reading off the radioactive bands, with a new set of values and practices.

These new values and practices were founded on the idea that manual sequencing, and the resulting autoradiographs, were 'inelegant' for automation.[22] The Caltech team was using the idea of *elegance* in a different way than Coulson had when explaining the pervasiveness of Sanger's techniques. With elegance, Coulson referred to the ability of both the plus and minus and dideoxy methods to mimic the natural replication process of the cell (see Chapter 2). The Hunkapillers and Smith sought an alternative model of elegance in a technological rather than a biological process: the assimilation of strings of data by the computer, to which they adapted their automatic sequencing technique.

5.5   String decoding and the computer as model

The new automatic approach to sequencing emerged between 1982 and 1983, after a series of informal meetings between Hood, Smith and the Hunkapiller brothers. Building on his doctoral research, Smith proposed attaching fluorescent molecules to the DNA and using a laser to detect them. Fluorescent, rather than radioactive, labelling had been used previously by both Hood's team and Smith's group at Stanford to visualise, respectively, antibodies and membrane molecules (Smith et al., 1980, 1981; Caltech, 1974–83).[23] T. Hunkapiller presented to meeting attendees an article which had recently been published by the journal *High Technology*. It described a chromatography method in which different molecules were separated into a thin tube filled with silicon. The Caltech researchers considered the possibility of applying this separation system to the labelled DNA fragments.[24]

Two different sequencing strategies were postulated as a result of these meetings, both based on Sanger's dideoxy technique (see Figures 2.3a,b).[25] The first one consisted of labelling the DNA fragments with the same fluorescent molecule and separating them in four lanes, as in manual sequencing. In the second one, a fluorescent dye of a different colour would be used in each of the four dideoxy reactions, so that the resulting overlapping DNA fragments would be distinctively labelled according to the nucleotide in which they ended: blue–adenine, green–cytosine, yellow–guanine and red–thymine. This would enable the separation of fragments in a single gel lane – as in the silicon tube – and produce a linear colour pattern corresponding with the DNA sequence. M. Hunkapiller was a strong advocate of this latter approach since, in his opinion, it would avoid most of the problems associated with automatic sequence processing (Cook-Deegan, 1994, pp. 65–8).[26]

The four-dye approach was adopted as the basis for a formal automation project by Hood's group. It was led by Smith, who became the head of a team of researchers and technicians specifically hired for the project. In 1983, M. Hunkapiller left Caltech to become a manager of ABI, easing cooperation between Smith's group and a team of biologists, chemists and engineers based at the company.[27] The researchers focused their efforts on strategies for attaching the colour dyes to the DNA fragments and detecting the resulting pattern with a laser. To this end, Smith devised a method for labelling the primers after which the different dideoxy fragments were formed following

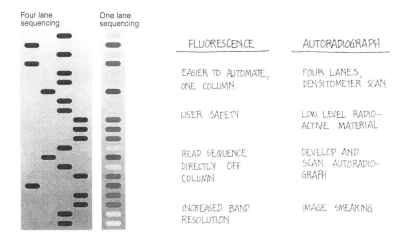

Figure 5.3a   The *autoradiograph world* v. the four colour dyes separated in one gel lane (left; in the original, the bands under 'One lane sequencing' are labelled in yellow, green, blue and red, whereas those under 'Four lane sequencing' are just labelled in black). Beside them, a 1984 report in which the perceived advantages of fluorescence are stated. (ABI Archives, Foster City, US, Steven Fung's Collection. Reproduced with permission.)

the action of polymerase (see Figure 5.3b). He also designed software to translate the laser stimulation of the dyes into sequence data.

In 1985, the Caltech team designed a prototype sequenator in which a laser beam located at the bottom of a gel lane detected the different colour labels of the DNA fragments as they moved down following separation. Information on their wavelength spectra was then transferred to a computer, which used software to determine the corresponding nucleotide sequence. In the annual laboratory report, Smith described the prototype as follows:

> This instrument [the sequenator] uses laser excitation of fluorescence to detect bands of DNA present in polyacrylamide gels. The DNA fragments generated in the enzymatic sequencing reaction are labelled with one of four fluorescent dyes, a different dye being used in each of the four different sequencing reactions. The four reactions are…co-electrophoresed in a polyacrylamide gel. The fluorescence detector measures the fluorescence at each of four different wavelengths, and this information is stored in a microcomputer. The DNA sequence is inferred from the temporal sequence of DNA bands. (Caltech, 1986, pp. 73–4)

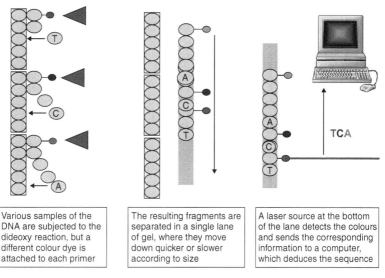

| Various samples of the DNA are subjected to the dideoxy reaction, but a different colour dye is attached to each primer | The resulting fragments are separated in a single lane of gel, where they move down quicker or slower according to size | A laser source at the bottom of the lane detects the colours and sends the corresponding information to a computer, which deduces the sequence |

Figure 5.3b The automatic DNA sequencing technique (small circles are coloured red, blue and green in the original)

Elaborated by author

All the modifications the Caltech researchers introduced into their new approach were intended to adapt the sequencing outcome to the requirements of the computer. During the early and mid-1980s, microcomputers had been implemented at Hood's laboratory and Smith was also familiar with this technology, as he had been developing software for laser detection in his previous research at Stanford. T. Hunkapiller, the main advocate in Hood's group for the introduction of computers, had been reviewing sequencing software and designing improved sequence analysis algorithms. At that time, Staden, Gingeras and Roberts, among others, were devising software adapted to microcomputers based on string manipulation algorithms commonly used in other text editing applications, such as operating systems and word processors (see Part II).

Despite the sequencing software being based on string processing algorithms, Sanger's techniques did not produce a string as an outcome. Due to computers being introduced into his group after development of his techniques, Sanger's sequencing emerged as an entirely chemical and biological form of work. His sequencing methods allowed researchers to read off DNA sequences, but the band pattern from which they could be read was spread over four autoradiograph lanes. This forced researchers to either manually input the sequence into the computer, or to use an interface which transformed the two-dimensional band pattern into a string. The designers of such interfaces were the members of the emerging sequencing software community and held different expertise from Sanger and the other inventors of manual sequencing.

The new automatic approach developed at Caltech transformed the sequencing outcome into a string. The microcomputer, as an information technology increasingly used by Hood's group, became the model for the automation of sequencing in a process mirroring the telephone and its role in the development of the area code hypothesis. In the mid-1980s, a decade after that hypothesis had been formulated, the computing practice of string production was incorporated into sequencing as a form of work. Sequencing was no longer a chemical and biological technique adapted to the computer through interfaces. The Caltech approach produced a string of colours which could be processed by microcomputers without further human mediation.

This automatic processing of the colour string by the sequenator was enabled by the addition of a laser. Laser technology had emerged in the 1950s with a Cold War military focus and was predominantly used as a weapon, in telecommunications or as a radar device. In the following decade, it was increasingly applied to the improvement of materials, environmental science, chemistry and medical diagnosis (Bromberg, 1991, pp. 208ff.). In this latter context, practitioners increasingly submitted their patients to laser exposure in order to destroy or detect diseased tissues, such as tumours, which had often been labelled with fluorescent molecules (Deutsch, 1988). Smith's work at Stanford on protein and lipid detection in membranes was an adaptation of this medical usage.

During the 1970s, lasers began to be used by the computer and software industries as detectors of digital information. Laser diodes were applied to 'read digital data' stored on computer discs. The data was codified in a succession of 'spots about 1 micrometer in diameter' and the reflection of the beam on each of the spots created 'readable information', which could be transmitted to a computer (Guenther et al., 1991, pp. 232–3). At the time of Smith's arrival at Caltech, this use of lasers was spreading among a series of instruments used in daily life, such as CDs or product-identifying barcodes in supermarkets (Campbell-Kelly and Aspray, 1996, pp. 161–2).

The transformation of the DNA sequence into a colour string fostered the use of the laser as an information reader by Smith's team. The prototype sequenator directed the laser to the detection of digital entities – a linear succession of fluorescent dyes – which were transformed into computer storable sequence data. A 1984 report resulting from the cooperation between the Caltech group and ABI defined the part of the sequenator where the laser acted as the 'read region'.[28] This suggests that the laser was being incorporated as an information technology, rather than as a detector of biomedical molecules, as in Smith's previous research.

The introduction of the laser adapted the whole sequencing process to the practices of string production and processing, which originated in the computer and software industries, and then spread among biological laboratories. Caltech's sequenator represented a strategy in which the chemical and biological steps of Sanger's sequencing had been altered in order to facilitate the handling of data by a microcomputer. Firstly, the sequencing outcome had been transformed into a digital string and, secondly, a laser beam read and transmitted the sequence information to be stored on a computer disc.

This adaptation of sequencing shows that the incorporation of computers into biology not only transformed the available string manipulation algorithms and database models, as in the cases of sequencing software and databanks (see Part II); it also led biological processes to be increasingly modelled on the categories and practices of the computer and software industries.[29] The increasingly computerised framing of sequencing resulted in an alternative form of work, originating in Hood's group, which substituted the practice of reading off the sequence with string coding and decoding. During the second half of the 1980s, this new computerised incarnation of sequencing gradually superseded those based on Sanger's and Gilbert's techniques.

5.6   Rival automation attempts

The emergence of Caltech's prototype sequenator coincided with other automation efforts in Europe and Asia. These efforts, in line with Sanger and Gilbert's techniques, produced a two-dimensional band pattern spread in four lanes as a sequencing outcome. The computer processing of the sequence thus required further manipulation in order to transform the band pattern into a string. Some of these automation efforts resulted in commercial sequencers which competed with Caltech and ABI's during the late 1980s. The preference of one sequencer over the others was deeply rooted in the contrasting values and attitudes attached to Sanger and Hood's group sequencing approaches.

Akiyoshi Wada at Tokyo University sought to automate the interpretation of the black and white radioactive band pattern of the autoradiograph, and produced an apparatus with support from a conglomerate of companies – among them Fuji – and the emerging biotechnology industry of Japan (Wada, 1984, 1987). Wada was the main representative of an effort by the Japanese Government to foster biotechnology and automatic DNA sequencing in the country. During the mid-1980s, Japan became one of the most serious biotechnological competitors of the US, prompting large-scale investments and policies by Ronald Reagan's Administration designed to boost sequencing and biotechnology entrepreneurship. Caltech's sequenator, ABI, and the HGP, were all direct beneficiaries of such initiatives (Cook-Deegan, 1994, pp. 70–1, ch. 15; see Chapter 6).

At the EMBL, the development of sequencing was significantly aided by reform of the Instrumentation Division during the early 1980s. Thanks to a boost to biochemical techniques within the Division (see Chapter 4), Wilhelm Ansorge, member of a group focusing on separation techniques, transformed his project on the optimisation of gels into an automatic sequenator (European Molecular Biology Laboratory, 1981, pp. 95–6). His apparatus, like Caltech's prototype, incorporated fluorescent labelling of DNA fragments resulting from the application of manual sequencing. However, rather than introducing colour distinction according to the nucleotide in which each fragment ended, he used the same fluorescent dye for all DNA fragments. The result was therefore a band pattern which maintained the structural features of the autoradiograph: it was monochromatic with the bands spread over four lanes (Ansorge et al., 1986, 1987).

Ansorge's sequenator was commercialised by the Swedish pharmaceutical multinational Pharmacia during the late 1980s.[30] It had some initial acceptance among the researchers who had previously been involved with Sanger and Gilbert's techniques – researchers used to the interpretation of two-dimensional band patterns and, crucially, more confident with these sort of procedures than with Caltech's re-elaboration of sequencing (see Chapter 6). Nevertheless, Caltech's approach was gradually established as the standard for automatic DNA sequencing.

In analysing Caltech's triumph over rival attempts, sociologists Peter Keating, Camille Limoges and Alberto Cambrosio have argued that 'successful automation' does not always imply 'automated mimicking of human operations'. The object of Caltech's later automation effort was not the activity previously performed manually by the researcher – the preparation and interpretation of an autoradiograph – but a new sort of procedure: the production of a colour string to be processed by a laser. This new sort of procedure, according to the authors, redistributed actions 'between humans and machines and between the humans themselves' (Keating et al., 1999, p. 132).

This conclusion is based on Keating, Limoges and Cambrosio's distinction between 'automation in the laboratory' and 'automation of the laboratory.' In the former, certain parts of a laboratory process are automated, with the process continuing to be performed as it had been manually. In the latter, the object of automation is the whole process and, therefore, the way of conducting it is transformed (ibid., pp. 127–8). Caltech's effort clearly squares with the latter and explains the redistribution of the practices of sequencing around the computer, adopted as the model for automation by Hood, the Hunkapiller brothers and Smith.

The redistribution of sequencing created a new form of work embodied in Caltech's prototype sequenator and organised around the practices of computer string processing. This new form of work squared with the institutional and personal values that dominated Hood's group and in many instances challenged those at the LMB. The sequenator eliminated human intervention in the determination of the DNA sequence, something inconceivable by Sanger's group. It was designed as an apparatus to be commercialised by a private company and became an important source of revenue for Hood's group and Caltech.

The opposition between the sequenator and some of the values and practices behind Sanger's techniques contradicts some accounts in which the automation of sequencing seems a logical and almost inevitable extension of the previous manual methods (e.g. Judson, 1992; Cook-Deegan, 1994, ch. 4). The form of work embodied in manual sequencing coexisted with the sequenator during the second half of the 1980s and, in some instances, was presented as an alternative to Caltech's approach. The researchers who were critical of Caltech proposed an automation model based on collaborative networks and scanning devices directed at the autoradiograph. They linked this model to the values of scientific accuracy, human control and cooperation over competition. The final rise of the sequenator was to a large extent a consequence of how ABI transformed the prototype into a commodity, and organised its marketing and commercialisation.[31]

6
The Commercialisation of the DNA Sequencer

The DNA sequenator was not the first instrument to be transformed into a commercial product by ABI. The company had previously launched a protein sequencer and synthesiser from Caltech's prototypes, as well as a DNA synthesiser between 1982 and 1983 (Applied Biosystems, 1983, pp. 2–7 and 17). The history of ABI overlaps with that of a series of small companies – start-ups – which were created from university departments in the second half of the 1970s and early 1980s, mainly in the US. They were inspired by the perceived potential of recombinant DNA and attracted the interest of a group of venture capitalists who specialised in risk investments. The proliferation of start-ups led to the configuration of a new business field – the biotechnology market – which in the mid-1980s was regarded by biologists as an unprecedented instance of cooperation between academia and industry (e.g. Kenney, 1986). However, the development of the chemical and pharmaceutical industries at the turn of the twentieth century constitutes a clear antecedent.

This chapter will place the emergence of ABI in the context of the changing, long-term interactions between academia and industry in the US and Europe. It will explain why this company was established in the Bay Area of San Francisco and explore its connections with a previous industry which had emerged in this area, concerned with semiconductors and other computer components. The organisational skills and expertise derived from these connections, as I will demonstrate, were essential for both the success of ABI within the biotechnology market and the transformation of the prototype sequenator into a commercial DNA sequencer. This transformation involved managerial and marketing practices designed to overcome the initial reservations towards the sequencer by some molecular biologists. By the early to mid-1990s, ABI's sequencer was an almost universally accepted instrument and

Figure 6.1 ABI's automatic sequencer (above) and Caltech's prototype sequenator (below). (Copyright the Science Museum (London) and courtesy of Lloyd Smith. Reproduced with permission.)

contributed decisively to the transformation of sequencing into an entirely mechanised form of work, one directed towards increasingly ambitious projects. The Human Genome Project, along with similar initiatives started at that time, had a decisive impact on the triumph of automated and large-scale sequencing.

6.1   Academic–industrial complexes

At the time of ABI's foundation in 1981, the US had developed a historically specific model of interactions between public laboratories and private companies. The historiography of the chemical and pharmaceutical industries has documented long-lasting and changing interactions between newly created companies and academic biological institutions, both in the States and Europe, between the late nineteenth and early twentieth centuries. This cooperation, often controversial, fostered the emergence of joint biomedical programmes and strategic alliances which were consolidated during World War I and the interwar years. It also led a number of core companies – DuPont, Merck, Pfizer and Imperial Chemical Industries ICI, among others – to adopt innovative organisational capabilities, combining their commercial divisions with research and development departments working on the refinement or production of new drugs or chemical compounds (Swann, 1988; Chandler Jr., 2005; Quirke, 2007).

Historian Robert Bud has shown that the alliance between academic microbiologists and chemical companies in Germany resulted in a line of research on fermentation, from which the term *biotechnologie* was coined during the 1910s and 1920s. It referred to the use of microorganisms to transform certain substances and obtain commercially relevant products, from wine and milk to acetone. This early use of the term has led Bud to argue for a 'long-term history of biotechnology' and to challenge its later appropriation by molecular biologists, who have often presented biotechnology as a consequence of the emergence of recombinant techniques to alter the DNA molecule in the mid-1970s (Bud, 1998b). The extended tradition of contacts between biomedical laboratories and industry throughout the twentieth century – and thus largely preceding the development of recombinant techniques – reinforces Bud's argument.

This academic-industrial collaboration was boosted by the scientific policy adopted in the US during and after World War II. Large federal funds were then devoted to basic biomedical research and its transfer to industrial manufacturing and commercialisation encouraged. This fostered the mass production of penicillin during the war and the subsequent marketing, by pharmaceutical companies, of antibiotics derived from fermentation processes (id., 2009). Therefore, in the mid-1970s, when the opportunity of developing products from recombinant DNA arose, there were infrastructures in place for the commercialisation of research conducted at US universities. Caltech and Leroy Hood's group were particularly favourable cases, but other institutions – such as Stanford and Berkeley University, which created Genentech and the Cetus Corporation – were also considered major representatives of a first wave of biotechnology start-ups (Yi, 2008; Rabinow, 1996).

The emergence of biotechnology companies was decisively favoured by the election of Ronald Reagan as President of the US in 1981. During Reagan's first year in office, the Bayh–Dole Act was passed, which allowed universities and other public institutions to patent discoveries derived from research sponsored by the NIH. Exclusive licence to the commercial exploitation of these discoveries could then be granted, in exchange for royalties, to start-ups or large pharmaceutical companies that developed drugs or other commodities. This permitted the companies to manufacture and market products without conducting all the research previously necessary, and therefore benefit from work funded by the public sector (Angell, 2005, pp. 6–7; Wright, 1994; Gottweis, 1998a).

Reagan's legislation fostered the proliferation of start-up companies – among them ABI – and the emergence of a consolidated biotechnology market. This was partly a response to the boost the Japanese Government was giving to Wada's sequenator and other products which appeared to threaten US leadership in the new recombinant DNA technologies (Collins, 2004; Bud, 1993, pp. 199–202; Orsenigo, 1989; Office of Technology Assessment, 1984; see Chapter 5). Molecular biologists adapted to the new situation by favouring an 'applied', industry-related orientation of research, absent in the origins of their discipline (Gottweis, 1998b, pp. 107–14). The most active molecular biologists at that time were a new generation of researchers, whose entrepreneurial attitudes distinguished them from their mentors.

The formation of this new 'academic-industrial complex' led some biologists to compare biotechnology with the organisation of US research following World War II.[1] As Marcia Angell has noted, however, there were important differences. Firstly, before the Bayh-Dole Act the results of NIH-funded research were in the public domain and could be adopted by any company without the privilege of exclusive rights. Secondly, the new legislative framework 'transformed the ethos' of researchers and their institutions, which saw 'themselves as partners of industry' and 'became just as enthusiastic as any entrepreneur about the opportunities to parlay their discoveries into financial gain'. For US citizens in general, 'it became not only reputable to be wealthy, but something close to virtuous': there were winners and losers, 'and the winners were rich and deserved to be' (Angell, 2005, quotes from pp. 6 and 8).

These transformed social and scientific mentalities shaped the subsequent perception of biotechnology as the paradigm of academic-industrial cooperation. The US biotechnology market started to be considered as the natural evolution of this cooperation, rather than a particular and historically contingent configuration of the changing interactions between public biomedical laboratories and companies. Current academic research on innovation still falls into this perception when considering the US model as the referent against which technology transfer in Europe and elsewhere should be compared (e.g. Valentin and Jensen, 2004; Chataway, Tait and Wield, 2004). The prevalence of this model has also led both historians and businessmen to argue that the UK failed to develop biotechnology as an independent market, due to the conservative attitude of its industry and the excessively interventionist approach of the government's research councils (Dart, 1989; see also Gottweis, 1998b).

But the emergence of biotechnology might also be seen as a historical process in which economic spaces were reconfigured through different – and dynamic – regimes of public and private appropriation of new biomedical resources (Harvey and McMeekin, 2007; Parry, 2004, chs 2–3). In this process Britain, rather than being a latecomer to the US model, can be regarded as an economic space marked by the public control of private appropriation. In Britain, as in the US, the funding of biomedical research increased significantly after World War II. However, the interactions between academia and industry were marked by a strong orientation to public service in both publicly-funded laboratories and government bodies (Liebenau, 1987, 1989). This orientation explains the cautious attitude of the LMB and the MRC towards commercialisation of the research they conducted and funded: both institutions feared that private appropriation would prevent citizens from access to research results supported by taxpayers' money (see Chapter 5).

The resistance to private appropriation survived the neo-liberal shift in the UK after Margaret Thatcher's rise to power in 1979. During Thatcher's early years in office, a strong pro-business orientation was fostered within British state-funded science. Consequently, the regulation of the emerging biotechnology market tended to converge with the US and to favour commercialisation of public research results (Wright, 1998). The resulting start-up companies, however, maintained a strong degree of public control: they either derived from privatisation processes or were managed by the Government's research councils, most of whose staff had been appointed before Thatcher and did not necessarily endorse her plans.

In 1980 the MRC, despite its opposition to project-oriented funding, created Celltech, a biotechnology start-up with preferential rights to commercialise products derived from research at the LMB (Bud, 1993, pp. 205–6; Dogson, 1990; Balmer and Sharp, 1993). The company devised a DNA synthesiser before ABI but, rather than marketing it, concentrated its product catalogue on the tailored nucleotide sequences it produced. Like the LMB faculty, the Celltech managers considered that technology should satisfy particular research necessities rather than automate repetitive laboratory practices. Celltech also focused on monoclonal antibodies, a technology developed at the LMB, the lack of commercialisation of which had been labelled a 'scandal' by Thatcherites and other socio-political actors (Celltech, 1981–86; de Chadarevian, 2002, pp. 353–62; see Chapter 5). ABI, by contrast, commercialised its synthesiser as soon as it was ready and made instrument development its sign of identity within the emerging biotechnology market.

6.2   ABI as an instrument-making firm

The orientation of ABI towards instrumentation was partly a consequence of its geographical location in Foster City, close to San Francisco. The West Coast, and particularly the Bay Area in Northern California, was where the first US biotechnology start-ups emerged. The Bay Area was also the home of Silicon Valley, where a previous business explosion, involving integrated circuits and other computer components, had taken place during the 1960s and 1970s. The founders of these computing companies – among them National Semiconductor and Intel – became venture capitalists who sponsored the early biotechnology firms (Lécuyer, 2006, pp. 5–8). Over time, these pioneer biotechnology companies also invested in later ventures. ABI emerged in 1981 from venture capital provided by Genentech and other prime movers in the new biotechnology business (Applied Biosystems, 1983, back page).

Universities in the Bay Area – such as Berkeley or Stanford – were less reluctant than their Eastern US counterparts to create start-up companies to which, following the Bayh-Dole Act, they licensed the commercial exploitation of the achievements of their academic staff. Walter Gilbert, the co-inventor of manual sequencing, faced fierce criticism from his home institution, Harvard University, when in the early 1980s he created the biotechnology firm Biogen.[2] He based the general headquarters of the company in Switzerland rather than Massachusetts, and in 1982, Harvard's authorities forced Gilbert to leave his professorship, though he was readmitted two years later. Harvard had refused to create its own biotechnology start-up despite the favourable legislative framework (Mendelsohn, 1992, pp. 17–19).

Caltech did not allow its researchers to directly own companies but, as in most universities in California, they could easily engage in business activities by becoming external consultants (Cook-Deegan, 1994, pp. 67–8). This resulted in a proliferation of start-ups in this area and in strong competition between them. The many expected applications of recombinant DNA led some early biotechnology companies to set excessively ambitious goals: they intended to focus simultaneously on the production of drugs, chemicals, kits to diagnose hereditary diseases and computing tools to gather DNA sequence data.

However, as the 1980s went on, increasing turbulence shook the emerging biotechnology market. Private and public actors struggled over the proprietary control of the new biodata, particularly the possibility of patenting DNA sequences. Though the battle seemed initially favourable to the biotechnology companies, the proliferation of open-access DNA sequence repositories – such as the EMBL Data Library and Genbank – made the collection of bio-information commercially unattractive (Harvey and McMeekin, 2007, ch. 2; Bostanci and Calvert, 2008; McAfee, 2003). A number of medically-oriented start-ups and pharmaceutical companies then focused on the interpretation of the open-access databases (Parry, 2004, chs. 5–6), but that required investment in specialised software and intensive time-consuming efforts, the success of which was not guaranteed. Furthermore, the early biotechnology start-ups faced disappointing results in their first balances, leading them to reduce their risks and target safer markets in order to avoid bankruptcy (Bud, 1993, pp. 190–5; Rabinow, 1996, pp. 31–7, 46–53).

This experience led to the demarcation of ABI as a company strictly devoted to the manufacture of instruments and reagents. The firm was initially to be named GeneCo and its founders drafted a series of briefings prior to its creation. In them, they considered that 'few, if any, companies' were poised 'to focus exclusively on supplying equipment and supplies' to the emerging biotechnology industry. The new company would thus be formed 'specifically to capitalise on this opportunity' and to produce 'tools such as biochemical, synthetic, analytic, separation, and purification instruments.' Its planned clients would be the 'practitioners of genetic engineering, molecular biology and biochemistry'.[3]

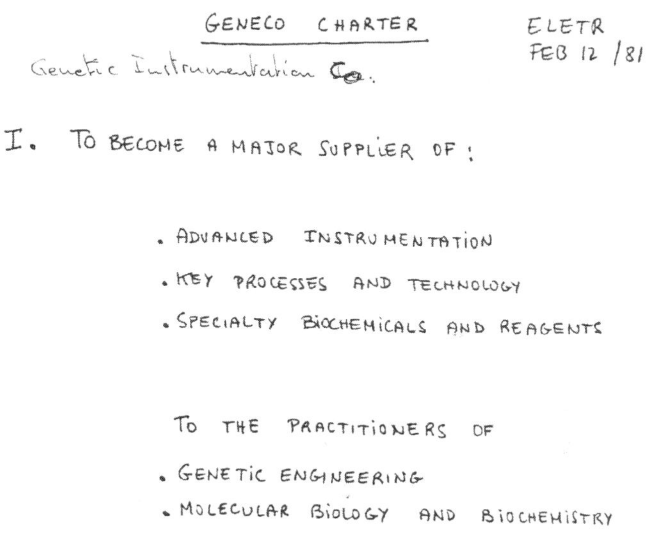

GENECO CHARTER

Genetic Instrumentation Co.

ELETR
FEB 12 /81

I. TO BECOME A MAJOR SUPPLiER OF :

· ADVANCED INSTRUMENTATION

· KEY PROCESSES AND TECHNOLOGY

· SPECIALTY BiOCHEMICALS AND REAGENTS

TO THE PRACTITIONERS OF

· GENETIC ENGINEERING

· MOLECULAR BiOLOGY AND BiOCHEMISTRY

Figure 6.2 A draft of goals and target customers written in 1981, when ABI was planned to be called GeneCo. The company was conceived as primarily devoted to biological instruments and reagents. (ABI Archives, Foster City, US, André Marion's Collection. Reproduced with permission.)

The first annual reports of ABI consolidated the company's focus on 'the most important and frequent laboratory activities' in biotechnology: 'protein and DNA sequencing' together with 'protein and DNA synthesis' (Applied Biosystems, 1983, p. 4). The firm aimed to manufacture and commercialise 'fine biochemicals and automated instruments' for 'biochemistry, molecular biology and biotechnology research and applications' (id., 1984, p. 1). The catalogues also advertised automatic instruments related to medical diagnosis. Over time, however, these instruments were not continued in the face of uncertainties and the saturation of the medical biotechnology market, in contrast to the increasing success of the sequencers.[4]

The universality that sequencing and synthesis were acquiring within biomedical research – both in academia and the biotechnology industry – meant that biochemists, molecular biologists and medical researchers would all potentially be interested in ABI's instruments. At the time of ABI's foundation, the bulk of biotechnology companies – either start-ups or specialised divisions created within multinationals – concentrated on the application of these same instruments to the design of therapeutic substances or kits that would react with the mutations of genetic diseases. This restricted their potential customers to those using biotechnology for medical purposes.

The focus on instrumentation thus provided ABI with a position in the biotechnology market and reduced competition. Rival companies saw automatic instruments as tools for their medical investigations, rather than products suitable for manufacture. Before the creation of ABI, Hood had offered a license for the protein sequenator to Beckman Instruments and AmGen, respectively an instrument-making multinational and a start-up. Both rejected the offer, despite having invested in other research initiatives by Hood's group and at Caltech in general. Beckman Instruments stated that it already had a commercial protein sequencer, whereas AmGen preferred to focus on drug development (Cook-Deegan, 1994, pp. 67–8).

6.3   Managerial practices and the sequencer as information technology

Another reason for ABI's instrument-making orientation was the background of its founders. The first two Presidents of the company – André Marion and Sam Eletr – had worked during the 1970s at Hewlett Packard (HP), a major company in Silicon Valley. Marion was a manager of engineering processes, while Eletr worked as an engineer in the division of medical instrumentation. When the venture capitalists who funded ABI proposed these two as chairs, Marion and Eletr decided to export their expertise in electronic instruments and become leaders in the manufacture of the new instrumentation for biological research.[5]

Marion and Eletr administered ABI as an information technology company. Retrospectively, the former has claimed the products and manufacturing processes of ABI were 'remarkably similar' to those at HP and, consequently, he could adapt 'many of the strategies' he had used in Silicon Valley.[6] During the 1980s, ABI's annual reports repeatedly presented the manufactured instruments as 'systems to automate information handling and processing for life sciences applications' (Applied Biosystems, 1983, p. 4; 1986, central triptych; 1987, p. 4).

Marion also introduced into ABI a managerial practice he had used at HP of forming divisions which each combined research, manufacturing and marketing, in order to avoid excessive compartmentalisation in the company. The DNA Sequencing Division was constituted by three interrelated teams, with a leader and three to twelve working staff. The

Research and Development Team integrated chemists, physicists and biochemists subdivided into groups, and in charge of refining the DNA sequencing reactions and application of the laser. The Engineering Team, consisting of electronic and mechanical engineers, was charged with building the actual apparatus. The Product Line Team included marketing agents who sought to make the sequencer attractive and easy to use by the customer. The three teams converged around the DNA sequencer, which was conceived by Marion as a 'system' which began 'in the first manufacturer' and finished in the 'final user'.[7]

The presence of engineers and, especially, marketing experts made ABI's division different from Hood's group at Caltech. ABI was devising a commercial product rather than a prototype to be used by a small group of researchers. The design of the product was essential for its success; the engineers and marketing agents were responsible for adapting the sequencing process into a functional and user-friendly apparatus. With this in mind, Marion formed customer support teams which supervised the manufacture processes of the DNA sequencer and the instruments produced at other divisions of ABI. Once the instruments had been commercialised, these teams solved the users' problems in their own home institutions and organised training courses to familiarise customers with their new devices (Applied Biosystems, 1983, p. 9).[8]

The formation of hybrid teams combining research, engineering and marketing was a managerial novelty which had been introduced in the 1960s. At that time, a number of companies challenged the highly functionalised models which had dominated business organisation since the end of World War II (Chandler Jr., 1977, pp. 1–12). Business historian Christophe Lécuyer has shown how Fairchild Semiconductor in Silicon Valley pioneered managerial strategies which 'closely coupled product development with market demands.' This combination of manufacturing and marketing skills was refined during the 1970s by the integrated circuit companies Intel and National Semiconductor (Lécuyer, 2006, pp. 165 and 292–4).

The incorporation of these managerial strategies by ABI shows that the shaping power of the computer went beyond the development of the prototype sequenator. The practices of the information technology industry affected not only the automation approach adopted at Hood's group, but also the organisation of the manufacturing of ABI's commercial sequencer. These transfers of practices and strategies between Silicon Valley and the biotechnology market were also essential in the further spread of automatic sequencing. ABI's customer support teams

1987 CUSTOMER TRAINING COURSE SCHEDULE

	Course No.		Date	Location
*	101	Protein Sequencer	January 13-16	Foster City, CA
*	102	Peptide Synthesizer	January 20-23	Foster City, CA
	103	Protein Sequencer	February 17-20	Foster City, CA
	104	Peptide Synthesizer	February 24-27	Foster City, CA
	105	DNA Synthesizer	March 2-4	Foster City, CA
	106	DNA Extractor	March 9-10	Foster City, CA
	107	DNA Sequencer	March 11-13	Foster City, CA
	108	Protein Sequencer	March 17-20	Foster City, CA
**	109	Peptide Synthesizer	March 24-27	Ramsey, NJ
	110	DNA Synthesizer	April 27-29	Foster City, CA
	111	DNA Extractor	May 4-5	Foster City, CA
	112	DNA Sequencer	May 6-8	Foster City, CA
	113	Protein Sequencer	May 12-15	Foster City, CA
	114	Peptide Synthesizer	May 19-22	Foster City, CA
**	115	Protein Sequencer	June 2-5	Ramsey, NJ
	116	DNA Synthesizer	July 8-10	Foster City, CA
	117	DNA Extractor	July 13-14	Foster City, CA
	118	DNA Sequencer	July 15-17	Foster City, CA

DNA SEQUENCER TRAINING COURSE

Description

The DNA Sequencer Course is a three-day session that will teach Model 370A operators how to obtain optimal results from their instruments. Topics covered include the recommended procedures for cloning and sequencing, software, troubleshooting and the synthesis of dye-primers.

Figure 6.3 ABI's training courses, organised by the customer support teams including, some devoted to the DNA sequencer (numbers 107, 112 and 118). Below, a description of the DNA sequencer courses. (ABI Archives, Foster City, US, Library Collection. Reproduced with permission.)

played a crucial role in making the sequencer acceptable to researchers accustomed to manual techniques.[9]

6.4   The reception of the sequencer and the different standards of sequencing

A major topic of discussion among biologists, engineers and marketing experts in ABI's DNA Sequencing Division was the representation of data by the automatic sequencer. Steven Fung, a Research and Development Team member in this Division, recalls that they considered a number of different formats in which the apparatus could present the DNA sequence output to the user. The team finally settled on a graph in which a series of wavelength peaks corresponded with the different colours of the dye-labelled DNA fragments. The peaks were the result of the detection of the dyes as the fragments moved down the gel and passed the laser beam, with the computer automatically deducing the sequence. This 'peak view' of the sequence was employed in ABI's first commercial DNA sequencer – the 370A model – released in 1986 (see Figures 6.1 and 6.4).[10]

The *peak view* contrasted with the outcome of both Sanger and Gilbert's manual methods, and Ansorge's rival automatic sequenator, to be commercialised by Pharmacia. The sequencing procedures in which these technologies were based required the separation of the DNA fragments into four lanes and therefore generated a two-dimensional band pattern: an autoradiograph of black radioactive bands in the case of Sanger and Gilbert, and a pattern of bands labelled with the same colour by Ansorge's apparatus (see Chapter 5). The ABI Division was aware of this difference and, in the instruction manual of the 370A, advised that both the sequencer, and the resulting data, should not be interpreted as previous representations of the sequence had been:

> Do not be confused! The Model 370A DNA Sequencer is a real-time electrophoresis detector. It should NOT be thought of in terms of the standard sequencing gel or gel reader. With the 370A, the bands (reaction products) are detected with a laser within a single (optimal) horizontal plane on the gel, AS THEY PASS THAT POINT. The information is stored by computer. At the end of the run, the data collected will be displayed on the screen and analysed. Note that it is not necessary (...) to be concerned about the separation of all the bands after a certain period of time. It is not necessary to save the gel as a 'picture' of your sequence information.[11]

These contrasting formats and operation shaped the reception of the sequencer in the early years of its commercialisation. Whereas a number of researchers embraced the new apparatus, others – surprising the ABI's managers – were unhappy with the 370A. Reservations were especially intense among biologists who had previously been involved with the manual techniques. These researchers were not used to interpreting a succession of wavelength peaks and felt more familiar with a two-dimensional band pattern. Technically, the 370A could have displayed a *picture* of the gel where single or various colour strings representing the labelled DNA fragments of single or various simultaneously determined

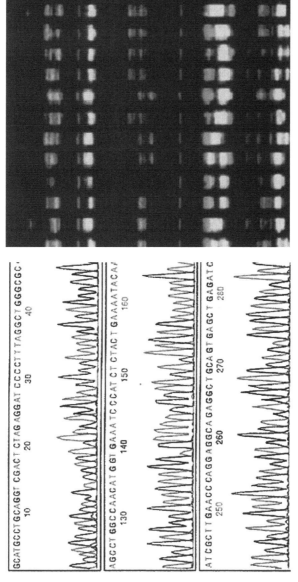

Figure 6.4 The peak view v. the gel view of the DNA sequence (above left and right). In the latter, each gel lane represents a different sequence – a complete set of fluorescently labelled DNA fragments – simultaneously determined by the sequencer. Both the peak lines and bands are labelled in different colours in the original. (ABI Archives, Foster City, US, Library Collection. Reproduced with permission.)

sequences separated. However, this possibility – named a gel view – was ruled out by ABI, which considered peaks the most convenient and coherent representation of the data within Caltech's new approach to sequencing.[12]

ABI's sequencer and its uneven reception were partly framed by the different values and attitudes of researchers towards automation. During the late 1980s, as the 370A was being commercialised, a number of automatic autoradiograph readers, for both DNA mapping and sequencing, were launched, among them one based on the LMB's software and marketed by Amersham (Gee, 2007, ch. 4). Amersham had been the main supplier of radioactive isotopes for Sanger's manual techniques (see Chapter 1) and was privatised during Britain's move towards biotechnology entrepreneurship in the early 1980s.

One of the alleged advantages of Amersham's reader was the possibility of a human check of the data against the autoradiograph, an option which was not available with the peak view of the 370A. Equally, competing sequencers developed by other companies incorporated a representation of the results closer to the gel view, often stating in their publicity that the sequencers 'spoke the same language as researchers' (Keating et al., 1999, pp. 131–3).

Another reservation about the 370A was that ABI, a private company, would control the interpretation of the sequence data and the lucrative market of reagents – gels, dye-labelled fragments or polymerase – if its sequencing conditions were established as standard. This control seemed inappropriate to researchers defending the public orientation of sequencing and universal availability of results, among them those based at the LMB and other institutions which shared its *culture*. Conversely, other biologists enthusiastically embraced the 370A model without excessively worrying about these matters.

A major defender of ABI's sequencer was Craig Venter, a young and ambitious neurobiologist at the NIH National Institute of Neurological Disorders and Strokes. During the late 1980s, he proposed sequencing hundreds of brain receptor genes involved in diseases, using the 370A. Venter became increasingly frustrated with the obstacles that, in his view, the NIH posed to automation and a large-scale sequencing initiative. He abandoned his home institution in 1992 and founded The Institute for Genomic Research (TIGR), a private not-for-profit organisation (Venter, 2007).

The acceptance of ABI's sequencer among the LMB Faculty was slower and more problematic. In 1989, John Sulston and his co-workers, Alan Coulson and Robert Waterston, initiated a 'world tour' of automatic sequencers to gain knowledge and experience, and consider their use in the recently approved *C. elegans* sequencing project. They visited the laboratories of Venter and Ansorge, among other institutions, and tested both ABI and Pharmacia's sequencers. The *C. elegans* researchers decided to purchase one of each machine and operated them simultaneously during the early 1990s. Finally, they chose ABI's version and used it intensely in the sequencing of *C. elegans* and the initial stages of the HGP, conducted at the Sanger Centre and Washington University in St Louis (Sulston and Ferry, 2002, pp. 85ff.).

Nevertheless, Sulston was unhappy with the automatic interpretation of the sequence by the 370A and decided to alter the apparatus. In his memoirs, he confesses that, shortly after its purchase, he decrypted the algorithms of the sequencer's software in order to obtain data about the gel positions of the DNA fragments. This allowed Sulston to regain control over the process, since the determination of the sequence could be done with software designed at the LMB:

> I could not accept that we should be dependent on a commercial company for the handling and assembly of the data we were producing. The company even had ambitions to take control of the analysis of the sequence, which was ridiculous. (...) So, one hot Summer Sunday afternoon, I sat on the lawn at home with printouts spread out all around me and decrypted the [ABI] file that stored the trace data. (...) I came in on Monday morning and said, 'Look, this is how we get the file data'. Within a very few days, [our programmers] had written display software that showed the traces [of the DNA fragments on the gel] – and there we were. (ibid., p. 94)

The criticisms of the 370A illustrate that during the late 1980s and the beginning of the 1990s, two alternative forms of work were coexisting and struggling to become the standard sequencing process. One introduced a certain degree of automation, but maintained human control over the interpretation of the sequence. It was represented by semi-automatic devices such as autoradiograph readers, always subordinated to the user. The other fully mechanised the determination of the sequence, and was embodied in ABI's sequencer. From a scientific viewpoint, this apparatus was autonomous and could be operated by technicians, separating biologists from direct engagement with the sequencing process.

A general preference for one or the other form of work was not apparent during the commercialisation of the 370A. The LMB researchers were well-respected and influential among the biomedical community, and many of them had been awarded the Nobel Prize. The Laboratory was still considered the *Mecca* of molecular biology and this discipline was seen by many as the direct antecedent and inspirer of biotechnology. In 1991, Bart Barrell, Sanger's former assistant and the new head of the sequencing division at the LMB, stated in a review paper that new technology could not just be 'bought with money'. Good ideas were needed, as well as people 'with the vision and determination' to make the technology work (Barrell, 1991, p. 44). That same year, one of the first popular accounts of the HGP raised doubts about the success of ABI's sequencer. The 'length of sequence' which could be determined with that apparatus was 'the same as with the original Sanger technique', which made the author wonder whether 'a well-organised team' equipped with manual procedures would be as efficient as 'one using the much more expensive' 370A model (Wills, 1991, p. 152).

These reservations led ABI to intensify their efforts to create a consensus among the community of users. During the early commercialisation of the 370A, the different teams of the DNA Sequencing Division debated strategies to improve the perception and user friendliness of the sequencer among its critics. The input of marketing agents and the customer support teams was crucial for a solution which, on technical grounds alone, appeared to be a backward step. The division, in the light of feedback from the training courses and troubleshooting reports, decided to devise a new generation of sequencer, incorporating some of the practices and representations of the rival form of work.[13]

The 373A model, released in 1990, produced the gel view as well as the peak view of the DNA sequence (Applied Biosystems, 1990b, p. 1). It represented a compromise solution, since the operation of the

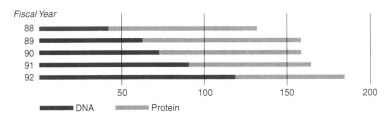

Figure 6.5 ABI's sales per product between 1988 and 1992. The DNA market (including sequencers) gradually increased and clearly overtook the protein market in 1992. (ABI Archives, Foster City, California, Library Collection. Reprinted with permission.)

sequencer maintained ABI's full automation of data production, but allowed researchers to check the position of the DNA fragments over the gel.[14] The response to the 373A was substantially more consensual than that of its antecedent, to the extent that this sequencer and other DNA-based products became ABI's most profitable merchandise during the early 1990s (id., 1990c). The company continued to face rival apparatuses, but even the alternative sequencers accepted the fully automated form of work embodied in the 373A.

6.5   Thermal cyclers and the introduction of PCR

Another area of interest for ABI during the development of the 373A was the automation of the dideoxy sequencing reactions. In the previous 370A model, researchers needed to conduct them manually in test tubes, prior to operation of the sequencer. Also, in 1990, the company launched a kit called the thermal cycler which, from a given DNA template, allowed the formation of the colour-coded and dye-labelled fragments with minimum human intervention (Applied Biosystems, 1990d).

The thermal cycler derived from an adaptation of the Polymerase Chain Reaction (PCR), a technique used to duplicate, almost without limit, a particular DNA sample. PCR had been developed at another biotechnology start-up in the Bay Area, Cetus Corporation, between 1983 and 1985. However, the medical orientation of Cetus, together with the professional and personal identity of the inventor, Kary Mullis, had prevented its prior introduction into sequencing or ABI's research and development activities.

PCR had originally been invented as a means of detecting the genetic mutation causing sickle-cell anaemia. Its conception and further refinement by Mullis was shaped by a main line of business development at Cetus: after setting excessively ambitious goals during its early days in the late 1970s, the company was gradually focusing on diagnostic tools for different hereditary diseases. Sickle-cell anaemia – a long-researched and well-known mutation system – was chosen as a model to develop a general diagnostic method. The role of PCR was to multiply the specific region where the mutation was located within the DNA contained in patients' blood samples. This multiplied region was then tested in order to check for the presence of the mutation (Mullis, 1990, 1998, ch. 1; Rabinow, 1996, chs 2–3; Saiki et al., 1985).

To achieve this, PCR incorporated similar tools as the dideoxy and plus and minus sequencing techniques: polymerase and primers.

However, instead of letting polymerase perform the duplication activity – as Sanger had done with his DNA methods (see Chapter 2) – Mullis successively stopped and restarted the enzyme by heating and cooling the DNA samples. Heat and cold led to, respectively, the separation and reattachment of DNA's two strands. As polymerase and primers could only act on single-stranded DNA, the successive heating and cooling cycles allowed them to continuously repeat their action and, consequently, to multiply DNA exponentially.

Mullis has retrospectively framed the divergence between PCR and sequencing in the different experimental approaches of chemists and biologists, and his strong professional identification with chemistry. Chemists – like Mullis – preferred to control molecular reactions, whereas biologists let them run within natural living processes, such as DNA replication (Mullis, 1990, 1998, ch. 3). As with Sanger's protein and DNA methods (see Part I), different research environments and disciplinary affiliations – chemistry and biology, Cetus Corporation and the LMB – led PCR and sequencing to emerge as divergent forms of work.

The orientation of PCR towards diagnosis, together with Mullis's chemical approach, meant the technique went unnoticed by biologists involved in sequencing. This was reinforced by Mullis's role at Cetus, as the head of a laboratory which produced synthetic DNA sequences. The laboratory synthesised oligonucleotides – short DNA probes – to be used by other researchers at the company, including those working on sickle-cell anaemia. The fact that Mullis did not conduct his own research, together with his strong and eccentric personality, led to marginalisation and occasional rows with the company's investigators, usually from a more biological background. Mullis's isolation resulted in frequent frustrations and, in 1985, both *Science* and *Nature* rejected a single-authored paper in which he described PCR. The technique was considered too specific for a journal aimed at the general scientific community (García-Sancho, 2008, pp. 199ff.). Other papers, authored together with researchers from the company – and oriented towards PCR's diagnostic use – were accepted by these journals (e.g. Saiki et al., 1985).

Mullis's situation began to change in 1986, when he resubmitted a new version of his single-authored paper – now co-written with his assistant, Frederick Faloona – to a special issue of *Methods in Enzymology*. This journal was devoted to extended descriptions of protocols and techniques of interest within biomedical research. The special issue was edited by Raymond Wu, who in the late 1960s and 1970s had developed

DNA sequencing methods, prior to Sanger (see Chapter 2). Wu realised the potential of PCR for sequencing and other forms of work in molecular biology and advised James Watson to invite Mullis to one of the Symposia on Quantitative Biology he organised in Cold Spring Harbor. He was invited before the paper in Wu's special issue was published (Mullis and Faloona, 1987).

The Cold Spring Harbor Symposia had played a crucial role in the formation of the molecular biology community. They were attended by both the founders of the field and new leaders of the biotechnology industry. The symposium to which Mullis was invited in 1986 addressed the 'Molecular biology of *Homo Sapiens*' and followed on from earlier, small-scale meetings, in which the sequencing of the human genome had been debated. Mullis's presentation described PCR and discussed its applicability to a number of areas of molecular biology, including sequencing (Mullis et al., 1986). For the majority of attendees, it was the first time they had encountered the technique.

Mullis's presentation was attended by Hood and prompted a line of inquiry at ABI on the applicability of PCR to sequencing. It was framed by the attempts to improve the acceptability of the 370A, without renouncing the increasing mechanisation of sequencing. In 1990, ABI obtained a license for certain uses of PCR – among them the thermal cycler – and two years later, merged with the chemical instrument multinational Perkin-Elmer (Applied Biosystems, 1990a).[15]

Perkin-Elmer (PE) had reached an agreement with Cetus to commercialise PCR-related products and was allied with the pharmaceutical multinational Hoffmann-La Roche, which acquired the technique patent in 1992. By then, Mullis had resigned from Cetus, following increasing tensions and litigations with the company over PCR's patent and scientific credit (Mullis, 1998).[16] The merger of Perkin-Elmer and ABI – which resulted in PE Biosystems – reflected the increasing acquisition of biotechnology start-ups by chemical and pharmaceutical multinationals during the early 1990s (Bud, 1998). PE Biosystems became a division of Perkin-Elmer concerned with biotechnology-related applications. One of the main product lines it developed was different generations of thermal cyclers, which were either incorporated into sequencers or marketed as sequencing accessories.

The thermal cyclers allowed dideoxy reactions to be conducted over extended periods of time and without the manipulation of test tubes. They were accompanied by attempts at ABI to introduce a new system for the labelling of DNA templates, in which the different dye-colours were attached to the corresponding dideoxy nucleotides instead of the primers, as had been the case in the first prototype and commercial 370A sequencer (see Figure 5.3b). This reduced the number of dideoxy reactions necessary to form the labelled DNA fragments from four to one, and consolidated the leadership of ABI's products over rival sequencers that had been launched by the chemical multinational DuPont and Amersham (Prober et al., 1987; Gee, 2007, ch. 4).

The new labelling system, together with the incorporation of thermal cyclers, transformed sequencing into an almost entirely mechanised form of work. During the 1990s, this form of work gradually became routine, especially after the introduction of robotic techniques to automatically load and unload the DNA templates into sequencers.[17] The practice of fully automatic sequencing was decisively encouraged by the proliferation of large-scale sequencing initiatives, and the growing socio-political consensus surrounding them. These initiatives, commonly framed in the new field of genomics, made the organisation and efficiency of sequencing the main concern of their promoters.

6.6   Genomics and the organisation of large-scale sequencing

The development of ABI's automatic instruments coincided with the proliferation of proposals to sequence the DNA of increasingly large organisms. The possibility of sequencing the human genome, which had been postulated as a distant chimera by a number of researchers, was first officially discussed in 1985, with molecular biologists at the same time presenting projects designed to map and sequence the DNA of complex multi-cellular organisms (Sinsheimer, 1989). Sulston's physical map of *C. elegans* was part of these initiatives, along with the mapping and sequencing of both yeast and the bacterium *E. coli*. These projects had a shared aim: to characterise the 'complete genome' of the proposed organisms (Coulson and Sulston, 1986, p. 7821, Olson et al., 1986, p. 7826 and Kohara et al., 1987).

In 1987, geneticists Victor McKusick and Frank Ruddle proposed the term 'genomics' to refer to these initiatives and founded a new journal with that title.[18] They believed the multiplication of mapping and sequencing initiatives justified the establishment of a 'new discipline' and encouraged researchers to submit articles on the progress of their projects, as well as the development of associated technologies. The editorial of the first issue of *Genomics,* entitled 'A new discipline, a new name, a new journal', asserted that entire genome mapping and sequencing should be a research priority in the upcoming years:

> Mapping all expressed genes ... regardless of whether their function is known, sequencing these genes together with their introns, and sequencing out from these is seen by many as 'the way to go'. The ultimate map, the sequence, is seen as a rosetta stone from which the complexities of gene expression in development can be translated and the genetic mechanisms of disease interpreted. For the newly developing discipline of mapping/sequencing (including analysis of the information) we have adopted the term *Genomics*. (McKusick and Ruddle, 1987, p. 1)

Sequencing the complete genome of different organisms had been an objective among biologists since the dissemination of Sanger and Gilbert's techniques. Sanger himself, and other members of his group, spent the last years of the 1970s and the beginning of the 1980s determining the complete nucleotide sequence of a number of bacteriophage viruses and of human mitochondrial DNA (see Chapter 3). However, Ruddle and McKusick[19] were proposing the creation of a discipline around this endeavour, and attached to it promises of fundamental knowledge about development and the genetic causes of diseases.

The promises of genomics became more hyperbolic as the new field advanced. In 1992, two years after the HGP was officially launched, Hood announced a new approach to biology and medicine, focused on the collection and analysis of DNA sequences. This new approach contrasted with the traditional, hypothesis-driven way of conducting biomedical research and paved the way for a preventive and personalised medicine, based on periodical monitoring of the sequences of patients rather than the established model of reacting to disease (Hood, 1992). Gilbert predicted that sequencing would transform biomedicine into a 'theoretical' science, founded on computer-assisted analysis of data rather than conventional test tube experiments (Gilbert, 1992, p. 92).

This enthusiasm about genomics was not universally shared amongst the biomedical community. Some researchers, especially those outside mainstream molecular biology, argued against devoting so much time and effort to sequencing entire genomes at the expense of other areas of biomedical inquiry. They proposed a selective use of sequencing for analysing genes or, in other words, to restrict the sequencing efforts to the protein-coding regions of genomes (Hood and Smith, 1987, pp. 37–8; Brenner, 1990). Nevertheless, political agents and funding agencies were increasingly swayed by the promises of the proponent of genomics, at a time in which the personal computer and other information technologies were penetrating all realms of social activity. Therefore, a project aimed at collecting and analysing large volumes of DNA data with the aid of computers, software and databases had a good chance of obtaining funding and support.

Another reason for the triumph of genomics was its blessing by both old and new influential figures within molecular biology. Nobel Prize holders and the leaders of successful biotechnology companies used their scientific prestige to endorse the new field (Suárez-Díaz, 2010, pp. 67–72). The mapping and sequencing of whole genomes ensured both a new project to maintain their authority – social, scientific and political – and a market to sell the products of their companies. This lobbying activity was best illustrated by the increasingly intense campaign to sequence the human genome.

During the second half of the 1980s, human genome sequencing was debated in international scientific meetings and by government agencies of different countries. At the same time, molecular biologists announced their intentions to map and sequence the entire genome of different organisms, such as *C. elegans* or *E. coli*. In 1989, the co-discoverer of the double helix of DNA, James Watson, was appointed Director of the National Human Genome Research Institute of the NIH. He used his new position to persuade the proponents of the different mapping and sequencing projects to form an *international consortium*, and present their sequencing initiatives under a common label, as the HGP.

Historian and sociologist Michael Fortun has shown the difficulties in identifying the HGP as a single entity. He prefers to talk about 'the genomics project', in the form of a series of disperse mapping and sequencing initiatives which coalesced with the consolidation of the new field of genomics. This coalescence was crucially favoured by the strategic interests of different generations of molecular biologists. The common HGP label, and the promises it evoked, allowed them to secure unprecedented funding – both generous and long-term – from different national bodies, such as the US Congress and the Wellcome Trust. The success of this unitary label led the members of the HGP consortium to consider 1990 the 'origin' of the Project and subsequently build a narrative of progress around the aim of sequencing the human genome (Fortun, 1993, chs 1–2).

ABI benefitted directly from the HGP campaign, its subsequent narrative and long-term promises. In 1987, Hood and the designer of Caltech's prototype sequenator, Lloyd Smith, published a series of review papers in influential journals – among them the newly founded *Bio/Technology* (later *Nature Biotechnology*) – on how to accomplish the still hypothetical sequencing of the human genome. They concluded it would be unwise to directly embark on such an enterprise with current technology, proposing an initial focus on instrument development and model organism sequencing (Hood and Smith, 1987; Smith and Hood, 1987). This approach, which favoured the interests of Caltech and ABI, permeated the first official surveys on the feasibility of human genome sequencing. An influential 1988 report by the Office of Technology Assessment – a scientific advisory body of the US Government – made optimistic claims regarding the completion of the human genome, built on the assumption of a linear progression in sequencing technologies (Office of Technology Assessment, 1988).

In 1991, with the HGP already under way, Hood and other members of his Caltech team presented an organisational model to optimise the 'throughput' of DNA sequencers. By *throughput,* they meant the achievement of the maximum number of sequenced nucleotides per person and per time, with minimum cost. According to them, a high throughput was achieved by optimising the distribution of sequencers among sequencing centres, the number of operators in charge of them and the man-machine interactions. The authors expected that commercial sequencers would reach in 'the near future' one million nucleotide pairs ('Mb') per year, which would make sequencing projects of over 100 Mb 'feasible during the next 5 to 10 years' (Hunkapiller et al., 1991, quotes from p. 59). ABI's new generation of sequencers in the 1990s – the 377 and 3700 models – were presented as 'high throughput' technologies.[20]

This concern with laboratory organisation and high throughput suggests that, within the HGP framework, the optimisation of the sequencing process became more important than its day-to-day application. Fortun has defined genomics as 'faster' and 'more intense than genetics', arguing that increasing speed is its truly distinctive feature, when compared with the late 1970s and 1980s sequencing projects (Fortun, 1998, 1999, quote from 2008, p. 31). The importance of speed affected the distribution of labour in those centres involved in large-scale sequencing projects, resulting in the incorporation of managerial practices aimed at rationalising production and maximising productivity (Ramillon, 2007, chs 1–2; Balmer, 1996a,b, 1998). Researchers based in those centres have retrospectively invoked 'factory metaphors' to describe the new organisation of work (Hilgartner, 2004, pp. 116–18).

Not all the sequencing centres, however, organised their activity in the same way. Historians and STS scholars have questioned the adequacy of factory metaphors and assembly line models to capture the distribution of labour in some large-scale sequencing enterprises. They have suggested other organisational regimes and shown how, during the 1990s, factory-based proposals coexisted with alternative models, inspired by network cooperation rather than an assembly line (Bartlett, 2008; Hilgartner, 2008; Ramillon, 2008; Bonneuil and Gaudillière, 2007; Bonneuil and Thomas, 2009, ch. 10). The institutional cultures of the various sequencing centres, together with the attitudes towards automation of the researchers in charge of them, decisively shaped these different organisations.

A widely adopted model during the first half of the 1990s was to concentrate human and technological resources on the construction of a physical map of the genome, and then sequence the resulting DNA fragments. This model was assumed by the HGP international consortium and adapted from the organisation of the yeast and *C. elegans* projects, which had been established as pilots for the human genome initiative. The physical map allowed a certain degree of human involvement and control over the HGP, in line with the institutional cultures and values of some members of the consortium, such as Sulston. It permitted the automatically determined sequence to be checked against the assembled DNA fragments of the map.[21]

The organisation of labour at the consortium was challenged in 1998, with the emergence of a rival human sequencing initiative. Following the launch of the 3700 sequencer, Michael Hunkapiller, then President of PE Biosystems, persuaded Venter to create a private company, Celera Genomics, which was integrated into the same multinational group. Venter was then based in TIGR and had extensively used all the models of ABI's sequencers. This, together with the higher capacity of the 3700, led him to propose an alternative organisation of sequencing.

At Celera, the thousands of fragments into which the human genome was split were directly sequenced and assembled by overlap-detection software, without the necessity of reordering them against a prior physical map. This approach was intended to be faster than that of the HGP consortium; hence the name of the company – Celera means *swiftness* in Latin – and its slogan, 'discovery can't wait'. Venter named its sequencing strategy the shotgun method, a term which had been used by developers of the first DNA sequencing software – also directed at the detection of overlaps (see Chapter 3) – but with a different meaning, since they were not opposing any previous organisational model (Venter, 2007; Rabinow, 2005, ch. 1).

Celera and the international consortium competed over the accomplishment of the human sequence between 1998 and 2000. This race was marked by the consortium's intention to release the sequence into the public domain against Celera's wish to patent the data.[22] Following an agreement, both parties jointly announced the completion of the HGP in 2000, in an event hosted by British Prime Minister, Tony Blair, and US President, Bill Clinton. One year later, they simultaneously published their findings – without prior patenting – in different journals: Celera in *Science*, the international consortium in *Nature*. Philosopher Adam Bostanci has shown that the two sequences, despite being retrospectively inserted into the unified HGP narrative and corresponding to the same species – the three-billion nucleotide pairs of the human genome – represented different objects, the results of divergent values and practices (Bostanci, 2004). They embodied the rival sequencing strategies of Celera and the public consortium, shaped by their different histories and attitudes towards automation.

The sequences also embodied the general state of sequencing at the turn of the twenty-first century. The rival centres, despite their organisational discrepancies, were commonly engaged in large-scale enterprises performed by interconnected apparatuses in the hands of technicians. Their operations no longer required biochemical skills or dexterity at the laboratory bench, but involved a new set of practices, ranging from the maintenance of the sequencers to the operation of the associated software. The only space for the biologist in this new distribution of labour was to decide between different – often conflicting – management strategies.

This transformation of sequencing was the result of over half a century of history. From the mid-1940s onwards, sequencing had gradually shifted from a bench-based innovation in protein chemistry to a repetitive technique used in rationalised genomic centres and automated according to the parameters of the computer. This trajectory may seem obvious or self-evident when inserted into the progressive narratives of the HGP. However, when sequencing is addressed as a form of work, its development can be reconstructed without a pre-determined end. The long-term history of sequencing then emerges: a history that challenges accepted myths surrounding the recent development of biomedicine.

Conclusions: A Long History of Practices

> People ask why I want to get the human genome. Some suggest that the reason is that it would be a wonderful end to my career – start out with the double helix and end up with the human genome. That is a good story. It seems almost a miracle to me that fifty years ago we could have been so ignorant of the nature of the genetic material and now can imagine that we will have the complete genetic blueprint of man…. There is a greater degree of urgency among older scientists than among younger ones to do the human genome now. The younger scientists can work on their grants until they are bored and still get the genome before they die. But to me it is crucial that we get the human genome now rather than twenty years from now, because I might be dead then and I don't want to miss out on learning how life works.
>
> Watson, 1992, pp. 164–5.

In this book I have investigated the changing and sometimes conflicting identities that bio-molecular sequencing acquired during the second half of the twentieth century. Sequence determination emerged in the mid-1940s within protein chemistry and over the subsequent three decades it entered the technical and conceptual repertoire of evolutionary and molecular biology. In this transition, sequence determination absorbed a multiplicity of practices which were alien to its original institutional and disciplinary setting. Its employment within the study of evolution resulted in the adaptation of algorithms initially designed for the manipulation of strings of data in computer programming and, more generally, for information management in military centres, public administration and offices of multinational corporations. Usage within molecular biology led to the incorporation of instruments and experimental strategies designed to replicate biological processes rather than chemically degrade molecules. Molecular biology also framed the gradual expansion of sequence determination from proteins to nucleic acids, through the study of the so-called coding problem – that is, how DNA, as the genetic material, directed the synthesis of proteins.

RNA, and then DNA sequencing, emerged as a result of that expansion. From the mid-1970s onwards, sequencing was used in combination with a series of recombinant techniques designed to alter the structure of the DNA molecule, each of which had originated through different historical routes (Yi, 2008). Molecular biologists from different institutional cultures, and with differing scientific values and personal attitudes, proposed the application of sequencing to increasingly larger genomes; and they designed partially or fully automated instruments which could ease the practice of sequencing, and be commercialised. These automatic apparatuses represented a particularly successful business venture within the emerging biotechnology market and the technology on which the new discipline of genomics, as proposed in 1987, was founded. Both biotechnology and genomics were attached to prospects largely approved by political and funding agencies, promising fundamental knowledge about the mechanisms of hereditary disease from the analysis of DNA sequences.

The new sequencing instruments, disciplines and promises set the framework in which the first historical accounts of sequencing emerged. These accounts were largely written by two generations of molecular biologists who overlooked the longer history and presented sequencing as part of the so-called recombinant DNA revolution. According to these accounts, sequencing and other techniques to alter and determine the structure of genetic material paved the way for the promising biotechnology market and the new field of genomics. Likewise, recombinant techniques and sequencing were the result of increasingly sophisticated knowledge about DNA that molecular biology had provided, from the elucidation of the double helix in 1953 to decipherment of the genetic code. The history of sequencing could, thus, be inscribed as a story of progress in the concept of the gene, from an unidentified entity in the first half of the twentieth century, to a double helix of DNA in 1953 and finally, with the advent of genomics, a physical map and sequence (e.g. Watson, 2003; Judson, 1992).

These official histories portray sequencing as a technique of molecular biology, invented in the 1970s to be applied to entire genomes and leading, almost inevitably, to the Human Genome Project (HGP). They consider only the final identity of sequencing, ignoring its previous materialisations in protein chemistry and the study of evolution. At best, the previous sequence determination techniques and algorithms to analyse protein sequences are here presented as *remote antecedents*, invented by *precursors* of genomics or bioinformatics. But this teleological framework hides the historical specificities of sequencing, the institutional, generational and disciplinary transitions that led to its application to nucleic acids and then entire genomes, and the practices that persisted or changed in these transitions. Ignorance about these historical processes has resulted in a series of myths which define the current perception of sequencing as an essentially mechanised form of work, performed with automatic apparatuses in large-scale centres, and one which is transforming biology into a computer-intensive information science.

In this book I have presented a substantially different picture of the development of sequencing. The notion of a *form of work* – adapted from John Pickstone's scholarship (2000, 2007; see Introduction) – has constituted a helpful conceptual tool to overcome the official histories. Investigating when researchers began to sequence biological molecules, rather than retrospectively seeking the origins of the HGP, biotechnology or bioinformatics, allows us to determine the historically contingent emergence of a configuration of biomedical work which was subsequently transformed across different practices, actors, institutions and disciplinary interests. With this clearer perspective, it becomes evident that sequencing cannot be solely attributed to a technical revolution within molecular biology: it was a form of work which spanned various fields, practices and categories, such as science and technology, craft and automation, chemistry and biology, research and computation, pure and applied science or experiment and routine work. Indeed, its history demonstrates the inadequacies of these disciplinary and conceptual dichotomies for the analysis of knowledge production in the twentieth century (Edgerton, 1999, 2006).

Employing similar perspectives, historians are increasingly questioning the standard stories of genomics and other recent biosciences. The investigation of computer-assisted sequence analysis and the practices of collecting, comparing and computing protein sequences has demonstrated the long trajectory of sequencing and the necessity of a *longue durée* framework for its proper historical study. It has also opened the history of sequencing to new fields and actors, such as comparative biochemistry, molecular evolution and computer programming, as represented by Saul Needleman, Christian Wunsch, Walter Goad

Transitions	Time of occurrence	Resulting myth
From the design of experiments to the routine application of techniques.	Sanger's determination of increasingly large portions of the sequence of insulin (1940s and 1950s), and then viral DNAs (1970s and early 1980s).	The determination of bio-molecular sequences is the result of the application of experimental science (despite many of the laboratory staff in charge of sequencing today not knowing the experimental basis of the techniques in detail).
From the instruments, problems and strategies of protein chemistry to those of molecular biology.	Sanger's move to the Laboratory of Molecular Biology at Cambridge and application of sequence determination to RNA and DNA (1960s and 1970s).	Sequencing derives from molecular biology and an increasingly refined knowledge about the structure of DNA.
From test-tube experiments to computer-assisted analysis of data.	Sequence determination software and databases, initially for proteins, then for DNA (1960s to 1980s).	Biology is an information science focused on the analysis of *dry* sequence data.
From science conducted by academic biologists to service work offered by non-research professionals.	Incorporation of sequence database teams into biomedical centres (first half of the 1980s).	Sequencing is a revolutionary science (despite being mainly conducted by technicians).
From a human-led endeavour to an automated and highly rationalised process oriented towards large genomes.	Mechanisation of sequencing at Caltech and use of commercial sequencers in large-scale genomic projects (second half of the 1970s to 1990s).	Sequencing is a high throughput technique performed by automatic apparatus, as exemplified by the Human Genome Project.
From the fulfilment of individual and immediate goals to the accumulation of data for future (and not completely defined) use by other researchers.	Large-scale mapping and sequencing projects (second half of the 1980s onwards).	Large-scale mapping and sequencing centres are cutting-edge scientific institutions, in contrast to shared genomics and proteomics facilities at universities or research institutes, where small-scale sequencing is conducted on request.

Figure C1   The different identities of sequencing (elaborated by author) or the difficult career of Margaret Dayhoff (Suárez-Díaz, 2010, 2007; Suárez-Díaz and Anaya-Muñoz, 2008; Strasser, 2010, 2011). However, more work needs to be conducted, especially on the practice of gene mapping – and the role of human and medical geneticists in its development – the use and mechanisation of protein sequence determination, and the growing support of DNA sequencing, genomics and the HGP during the 1990s.

This growing professional historiography on sequencing may be relevant for ongoing regulatory debates in genomics and post-genomics. It illustrates the importance of history in the construction of the expectations and promises surrounding recent biomedicine. The role of the prominent founders of molecular biology and of a new generation of biotechnologists in proposing the HGP, as well as the presentation of this enterprise as a natural continuation of their research, suggest that they were projecting into the future a constructed history of their past.

From the mid-1960s onwards, molecular biologists – with increasing scientific and social authority – had written widely accepted autobiographies and historical accounts. This literature presented the double helix discovery as a watershed moment, one which provided genetics with a *hereditary substance* and marked the origins of molecular biology, portrayed as a discipline mainly focused on the investigation of DNA structure and function (Cairns et al., 1966; Watson, 1969; Jacob, 1974). These triumphant and epic stories constituted the background against which claims surrounding the HGP were formulated and gained sociopolitical momentum. If human genome sequencing was the natural continuation of progress within molecular biology, sequencing technologies would inevitably advance and allow an improved definition of the DNA molecule, leading to new and exciting achievements.[1]

A historical and critical perspective on recent biomedicine may also result in a necessary reassessment of what scientists currently mean by *genetic information*. This concept has had a crucial role in the experimental strategies and discourse of both molecular biology and genomics (Kay, 2000; Sarkar, 1996a,b; García-Sancho, 2007a,b). An approach to its history in light of the development of sequencing suggests the expectations around genomics may partly be derived from a conflation of different meanings of genetic information. For molecular biologists, genetic information was a set of instructions, external to DNA, which directed the synthesis of proteins. The emergence of the first DNA sequencing techniques in the mid to late 1970s led to a new meaning: genetic information became a determinable sequence of nucleotides (see Part II).

In his papers, Frederick Sanger, co-inventor of the techniques, described the act of determining the sequences as *reading off* the autoradiographs (Sanger and Coulson, 1975; Sanger et al., 1975; Sanger, Nicklen and Coulson, 1977). With this term, he referred to the extraction of sequence data from the pattern of radioactive bands which derived from the application of the techniques. However, in the subsequent decade, biologists began using the more general term *read*, especially when they addressed broader audiences. In 1987, the foundational manifesto of *Genomics* referred to the DNA sequence as a 'rosetta stone' and, the same year, Sanger described sequencing as 'reading the messages in the genes' (McKusick and Ruddle, 1987, p. 1; Sanger, 1987). The concept of genetic information as sequence data was thus being superposed on its previous meaning as a set of instructions. This suggested that, from DNA sequences alone, the genetic instructions – and therefore the function of the genes contained in the sequence – could be read and understood. The role of molecular biologists in the promotion of genomics, as well as the spread of the personal computer and other information-processing technologies at that time, may have contributed to this conflation of meanings.[2]

Lastly, the history of the introduction of software and database technologies into sequencing offers a perspective from which to critically reappraise the interactions of biology and computing. This perspective challenges the current view held by many bioinformaticians and some STS scholars, of a *natural* or *interdisciplinary* marriage between the two fields. During the 1960s and 1970s, as documented in the historiography of computing, a significant number of computational technology designers inhabited the administrative and commercial worlds, external to any academic discipline (e.g. Haigh, 2001; Campbell-Kelly, 2003). This suggests a traditional history of disciplinary interactions cannot fully capture how professional information managers were imported into academic biomedical laboratories, or the problems and consequences of this incorporation. Equally, the lack of communication between different biomedical fields created different routes for the introduction of computing, as shown by the discontinuities between the algorithms of protein sequence analysis and those of the first DNA sequencing software.

The histories of biology and computing have been characterised by their mutual shaping power, without one completely determining the other: computers are not merely neutral applications to biology, nor is current biology concerned exclusively with information management (Chow-White and García-Sancho, 2012).[3] The invention of Caltech's prototype sequenator illustrates how the introduction of computers had a decisive impact on the development of sequencing as a form of work and, more generally, on the way biology has subsequently been conducted. Conversely biological processes shaped the design of the first sequencing software and databases, which required a substantial re-elaboration of string processing algorithms in order to make them adaptable to the specificities of DNA sequences. Despite the sophistication of those algorithms, the diversity of biology has never been fully captured by computer programs or simulations. The emergence of biocomputing in the late 1980s has increasingly blurred the distinction between the biological and the computational, with researchers continually shifting from one to the other realm. Such movements question the currently fashionable separation between *wet* and *dry* biology, as presented in some STS research.

All these lines of inquiry suggest potential avenues for collaboration and a much needed methodological integration of the tools of history and contemporary STS.[4] They also clarify the puzzling sensations the visitor experiences when first entering the Wellcome Trust Genome Campus in Hinxton (UK). Its remote, rural location in Cambridgeshire is the result of the institutionalisation efforts of genomics and the necessity for independence from both the laboratory culture and the small-scale academic orientation of molecular biology, already consolidated at the University of Cambridge. The large-scale sequencing centre on the Genome Campus is the Sanger Institute, named after Frederick Sanger, whose extended career from proteins to DNA (1943–83) reflects the *longue durée* of this form of work. However, the names chosen for two important laboratory settings – the Morgan and Sulston Buildings – suggest an attempt to reconstruct the history of sequencing as a culmination of the accumulated progress of genetics, from the first chromosomal maps of *Drosophila* to the sequencing of the complete genome of *C. elegans* and then the HGP.

The general landscape of the Campus quickly counter-acts this attempted historical reconstruction. The combination of glass buildings with lakes and bucolic greenery reminds us that sequencing existed long before the Genome Campus, when Hinxton was a quiet, unfashionable village and the word *genomics* had not been invented. Even the older concepts of *gene* and *computer* would have sounded odd, not only to the Hinxton locals, but also to Frederick Sanger, who worked in nearby Cambridge and – despite inspiring the Sanger Institute – only sequenced DNA with computers during the last eight years of his career. The first DNA sequence database was initiated in faraway Heidelberg, only moving to Cambridgeshire after the foundation of the Genome Campus and the relocation of a small group of 'information engineers' to the glossy European Bioinformatics Institute. Determining sequences in the past was neither as DNA-focused nor as computer-intensive as is the case in Hinxton today.

The Genome Campus is, in sum, a materialisation of the current state of sequencing, combined with echoes of its long and often neglected history. It should, therefore, be seen as a particular embodiment of sequencing, rather than its unique or most convenient configuration. Sequencing as a form of work has historically preceded the disciplines, institutions and social formations into which it has been introduced. However, these disciplines, institutions and formations have been crucial for directing sequencing towards certain goals. The Genome Campus is the result of the introduction of sequencing into genomics and the increasing socio-political favour that this new field experienced during the late 1980s and 1990s. Within this Campus, sequencing is large-scale, fast and fully mechanised, but nothing prevents this form of work from travelling again in time or space.

The realisation of the historical trajectory of sequencing provides its users with the power to decide its destiny. Pickstone has noted that current biomedicine is characterised by the proliferation of analytical 'working knowledges', among them increasingly automated sequencing and bioinformatic techniques. These techniques may lead to a de-differentiation of disciplines, which increasingly base their research on sequences or other biomedical data (Pickstone, 2007, pp. 513–14; see also Leonelli, 2012). In these highly technified environments, researchers, regulators and stakeholders must carefully select the aims to which they direct the available automatic instruments and collected data.

It is possible that the rationalised and large-scale orientation of automatic DNA sequencing will be maintained. Genomic centres continue to sequence entire non-human DNAs, as well as those of patients with different hereditary diseases. Their claims on the future applicability of sequence data are intact, even if the promises of the HGP have not yet been fully accomplished. However, in the last decade (2000–10), sequencing has also been offered as a service in shared facilities within the many universities and biomedical institutions which did not participate in the invention of the techniques.[5] Within these facilities, sequencing is being reconfigured as a small-scale, slower and artisanal form of work, subordinated to concrete research necessities.

This reconfiguration leaves the future of sequencing in the hands of its users. The technicians, administrators and patron biologists of the shared facilities are demonstrating that, by regaining control over the technologies, other sequencing is possible. They can modulate the intensity with which the sequencers are operated and know in advance the questions that the resulting data will seek to answer. The usefulness and applicability of sequencing in these shared facilities is, thus, guaranteed without committing its future – and that of taxpayers' money – to the completion of a large genome.

Appendix I: Oral Histories

Name	Role	Location/date
F. Sanger	Inventor of the first protein and DNA sequencing techniques	Sanger Institute, Cambridgeshire, UK, Winter 2005
A. Coulson	Sanger's assistant during the development of the DNA techniques	MRC Laboratory of Molecular Biology, Cambridge, UK, Winter 2005
B. Barrell	Sanger's assistant during the development of the RNA and DNA techniques	Sanger Institute, Cambridgeshire, UK, Winter 2005
O. Kennard	Establishment of the Cambridge Structural Database	Private residence, Summer 2007
K. Murray	Initiation of the European Nucleotide Sequence Data Library	University of Edinburgh, UK, Winter 2007
G. Hamm	Development of the European Nucleotide Sequence Data Library	Phone interview, Spring 2007
G. Cameron	Development of the European Nucleotide Sequence Data Library	European Bioinformatics Institute, Cambridgeshire, UK, Spring 2007
J. Sulston	Development of the *C. elegans* mapping and sequencing project and Director of the Sanger Institute	Private residence, Summer 2005
L. Hood	Head of the Caltech group which developed a prototype automatic DNA sequenator	Institute for Systems Biology, Seattle, US, Spring 2006
T. Hunkapiller	Development of the DNA sequenator at Caltech	Institute for Systems Biology, Seattle, US, Spring 2006
L. Smith	Development of the DNA sequenator at Caltech	Westin Hotel, Seattle, US, Spring 2006
A. Marion	Co-founder of Applied Biosystems	Applied Biosystems, Foster City, US, Spring 2006
S. Fung	Member of the DNA Sequencing Division at Applied Biosystems	Applied Biosystems, Foster City, US, Spring 2006

Note: All reasonable efforts were made to obtain permission from the interviewees when their oral histories are directly quoted or referred to in this book. The unquoted interviews were used as general background for the investigation.

Appendix II: Archival Sources

Name	Location	Date of access
F. Sanger's laboratory notebooks	Wellcome Trust Archives and Manuscript Collection, London, UK (within the Archive of the UK Biochemical Society)	Spring 2008
Francis Crick's Papers	Wellcome Trust Archives and Manuscript Collection, London, UK	Winter 2005
Papers and Correspondence of A. Coulson	Wellcome Trust Archives and Manuscript Collection, London, UK	Winter 2006 (when still held at the Storage Building of the MRC Laboratory of Molecular Biology)
Archives of the MRC Laboratory of Molecular Biology	Storage Building of the MRC Laboratory of Molecular Biology, Cambridge, UK	Spring 2006
The National Archives of the UK	Kew, Surrey, UK	Winter and Spring 2005
G. Cameron's personal archive	European Bioinformatics Institute, Cambridgeshire, UK	Summer 2007 and Winter 2009
Papers and Correspondence of Sir John Kendrew	Bodleian Library, Oxford, UK	Summer 2011
Archives of the California Institute of Technology	Caltech Campus, Pasadena, US	Spring 2006
Archives of Applied Biosystems	Library and General Headquarters of Applied Biosystems, Foster City, US	Spring 2006

Note: All reasonable efforts were made to obtain permission from the archivists when the materials are directly quoted, referred to or reproduced in this book.

Notes

Introduction: An Historical Approach to Sequencing

1. A transcript of Clinton's speech at the HGP ceremony can be found at http://clinton5.nara.gov/WH/New/html/genome-20000626.html (last accessed July 2011). For a detailed analysis of the historical significance of the reference to a *Book of Life* see García-Sancho, 2007a.

2. Soraya de Chadarevian and Bruno Strasser have argued that proteins were a major research object during the first half of the twentieth century and had an important role in the formation of molecular biology, even after the elucidation of the double helix of DNA in 1953 (de Chadarevian, 2002, ch. 8; Strasser, 2002). This does not necessarily imply that DNA was unimportant or that the determination of its structure went unnoticed. A recent bibliometric investigation has demonstrated that the double helix paper was widely quoted shortly after its publication and opened new lines of research on DNA (Gingras, 2010).

3. This linearity can be taken even further back, to the foundation of genetics in the early twentieth century, prompted by the 'rediscovery' of Mendel's laws. Science journalist and historian, Horace F. Judson, has alleged that the 'fundamental discoveries of genetics' emerged as 'successive models of the processes of heredity'. Each model, he has argued, 'refined and made more complex the version preceding it.' Judson is therefore proposing a line of progress between the first geneticists – among them Morgan, the researcher after whom the Morgan Building at the Sanger Institute is named – and the HGP (Judson, 1992, p. 38).

4. STS scholars have also claimed the biological and the computational are increasingly indistinguishable in current societies. This has triggered a line of research on the cyborg and the so-called post-human body (Haraway, 1997; Hayles, 1999; Lenoir, 2002a,b; Zafra, 2010).

5. Protein sequencing has also been addressed by historians of biochemistry who, considering the DNA techniques to be part of molecular biology and, therefore, beyond their disciplinary scope, have resisted making the connection with DNA (Fruton, 1977, 1992, pp. 35–6).

6. The historicity of the interactions between biology and computing is more specifically addressed in a paper I co-authored with sociologist of biology Peter Chow-White (Chow-White and García-Sancho, 2012). On this pressing issue in the historiography of post-World War II biomedicine see also Lenoir, 1999; de Chadarevian, 2002, ch. 4, and November 2006. Another long-term historical perspective on sequencing is proposed in two collective volumes edited by Hans-Jörg Rheinberger and Jean-Paul Gaudillière. They portray sequencing as part of a number of interrelated 'mapping cultures' which shaped the development of twentieth-century biology. Other mapping cultures also addressed are, among others, genetic crossing, blood

grouping, and the investigation of developmental pathways (Rheinberger and Gaudillière, 2004a,b).

7. Gaudillière previously used the notions of 'styles of thought' and 'work' to investigate the history of sex steroids (Gaudillière, 2004a). Jonathan Harwood accounted for the differences between communities of German geneticists in their contrasting 'styles of scientific thought' (Harwood, 1993). The term *style of thought* was first coined in 1935 by Ludwig Fleck in his celebrated book *Genesis and Development of a Scientific Fact* (see Gaudillière, 2004a, pp. 526–7). Alistair Crombie and Ian Hacking have also distinguished different styles in scientific practice (Pickstone, 2011a, p. 235; id., 2007).

8. With the proposal of a two-directional dialogue I seek to avoid both technological determinism and an extreme form of social shaping, one which portrays technologies as inert instruments completely conditioned by their environment. On the interactions between technologies and society see MacKenzie and Wajcman, 1999; Edgerton, 1999.

Part I   Emergence: Frederick Sanger's Pioneering Techniques (1943–1977)

1. A more recent line of historical inquiry is exploring the role of protein sequencing in the conformation of comparative biochemistry and molecular evolution as distinct biological fields (Strasser, 2010; Hagen, 2010; Sommer, 2008; Strasser and de Chadarevian, 2011, pp. 321 ff.; Suárez-Díaz, 2009; Suárez-Díaz and Anaya-Muñoz, 2008). This literature will be discussed in Part II.

1   The Sequence of Insulin and the Configuration of a New Biochemical Form of Work (1943–1962)

1. Historian of biochemistry Joseph Fruton – who in his previous career as a biochemist worked with Bergmann – has claimed that the periodicity hypothesis was 'welcomed by some geneticists', since it set 'a mathematical principle that might link Mendel's [hereditary] laws to the structure of proteins' (Fruton, 1979, p. 13). The Rockefeller Institute had been a major setting for early investigations on the nature of genetic material, believed during the 1930s and 1940s to be protein. Phoebus Levene, former Director of the Institute's Laboratory of Chemistry and well connected with Fischer, proposed the tetranucleotide hypothesis in 1910, which – like the periodicity hypothesis for proteins – postulated a repetitive parameter in the structure of nucleic acids (Olby, 1992 [1974], pp. 81 ff.). This, together with the aspirations of the Rockefeller Foundation – owner of the Institute – to apply mathematical physics to biology, suggests the hypotheses of Bergmann and Levene were a product of their time rather than unfortunate *errors*, as some historians have stated.

2. The contrast between Bergmann and Niemann's mathematical approach to proteins and that of biochemists in Cambridge squares with a similar

dichotomy investigated by Lily Kay in the context of the coding problem. According to Kay, research on the way DNA synthesises proteins shifted from cryptographic methods to those of biochemistry during the 1950s and 1960s, in a similar fashion to events in the structural analysis of proteins twenty years earlier (Kay, 2000, chs. 3–6).

3. The number of insulin chains was a prominent research question at the time and the aim of Chibnall's initial request to Sanger (de Chadarevian, 1999, p. 203; Chibnall, quoted in Fruton, 1992, p. 36). In 1945, it was thought that insulin was composed of four instead of two chains. This was due to an estimation of its molecular weight that was higher than our present standards (Sanger, 1988, pp. 6–7; 1949b, pp. 157–9).

4. That Chibnall's group view of protein structure shaped Sanger's work has also been suggested by Fruton. Nevertheless, in 1952, Sanger still considered the unpredictable nature of proteins – and even the peptide theory – hypothetical (Fruton, 1992, pp. 43–4; 1999, pp. 215–18; 1972, p. 1).

5. Tuppy's career after Cambridge has been investigated by historian Bruno Strasser. It was based in Vienna and devoted to the determination of different protein sequences for further comparison (Strasser, 2010, pp. 627–9).

6. Sanger applied ionophoresis to his later insulin experiments and determination of RNA sequences during the late 1950s and 1960s. His first attempts with this technique are labelled as having been conducted 'in Uppsala' and bear the initials 'RLMS', referring to Richard [Laurence Millington] Synge (Sanger's Laboratory Notebooks, Wellcome Trust Archives, London, Insulin Book 3, p. 494 and Uppsala Book, reference number SA/BIO/P/1/6).

7. Quote from Sanger's Laboratory Notebooks, Wellcome Trust Archives, London, Insulin Books 4, p. 681. Despite only publishing one paper on gramicidin (Sanger, 1946), the first volumes of Sanger's notebooks – written between 1945 and 1949 – contain a number of experiments on this peptide scattered among those on insulin.

8. For more details on the development of analytical and synthetic chemistry see Chang and Jackson, 2007 and Pickstone, 2000, chs. 4–6.

9. Bergmann edited the last volumes of the collected works of Fischer, published in a series of eight issues, four of which appeared after his death (Fruton, 1985, footnote 30). One of them, offered by Fischer's son, accompanies Sanger's laboratory notebooks with the following dedication: 'To Dr. Fred Sanger in commemoration to his impressive presentation of his wonderful results in the Chemistry of Insulin' (Wellcome Trust Archives, London, item number SA/BIO/P/5).

10. One researcher pleased by this change of orientation was Pirie, Sanger's first PhD supervisor, who had complained during Hopkins's Chair about the lack of chemistry in the Institute's teaching programme (Kohler, 1982, p. 84).

11. The preference of Edman's technique over Sanger's was partially due to Edman's visit to the Rockefeller Institute and previous contacts with Stein and Moore. Edman himself was involved in the design of an automatic apparatus that could perform not only amino acid analysis, but all the steps of sequence determination. He named it the 'protein sequenator' and presented it in Australia, where Edman spent most of his later career (Edman and Begg, 1967). The development of Edman's technique and its role in the spread of protein sequence determination is a potential avenue of further

research. For more details on the automation of protein sequence determination see Part III of this book.

12. F. Sanger, interview with author, Sanger Institute, Hinxton, Cambridgeshire, UK, 2005. The only evidence of contact with crystallographers is a letter attached to the experiments in one of Sanger's insulin notebooks. In it, Hodgkin sends the result of X-ray analysis of a sample protein previously delivered by Sanger (Sanger's Laboratory Notebooks, Wellcome Trust Archives, London, Insulin Book 2, p. 232). The relationship between Sanger and Hodgkin, and more generally between crystallographic and chemical structure of proteins, deserves further research.

13. On Baldwin and the comparative tradition in biochemistry see Strasser, 2010 and Strasser and de Chadarevian, 2011. Other representatives of this comparative tradition were Anfinsen and Tuppy, respectively Sanger's persuader in the use of radioactive labelling and his assistant during the completion of insulin. Comparison between protein sequences would become an important practice in the 1960s within the emerging field of molecular evolution (see Chapter 3).

2 From Chemical Degradation to Biological Replication (1962–1977)

1. The hypothesis that DNA rather than proteins was the genetic material had been raised in the late 1940s by bacteriologists Oswald Avery, Colin MacLeod and Maclyn McCarty at the Rockefeller Institute, and was verified experimentally in 1952 by Alfred Hersey and Martha Chase (Olby, 1992 [1974], Section III). Historians have noted that these contributions have traditionally been obscured in retrospective accounts focusing on Watson and Crick (de Chadarevian, 2002, ch. 8). Watson and Crick did not refer to these investigations in their initial DNA papers.

2. Watson and Crick used the results of Erwin Chargaff at the University of Columbia and those of a group led by Maurice Wilkins at King's College London. Chargaff had published an analysis of the chemical composition of DNA, whereas Wilkins's group was applying crystallography to DNA and had already postulated a helical structure. Helical models were widespread in the investigation of biomolecules at that time and Linus Pauling, at the California Institute of Technology, had combined crystallography and chemistry to propose the α-helix model for the structure of haemoglobin. During the early 1950s, Pauling was working on DNA, adding competitive pressure. Scholars in gender studies and retrospective accounts by the involved researchers have suggested that Watson and Crick did not fully acknowledge the King's College group and Rosalind Franklin in particular, a pioneer in the crystallographic analysis of DNA (Maddox, 2003; Wilkins, 2005).

3. Another important cooperation for the investigations on haemoglobin at the Cavendish was that between Perutz, Ingram and Hermann Lehmann, a physician and medical researcher who collected haemoglobin variants from around the world and distributed them, upon request, among biologists and doctors (de Chadarevian, 1998a; Strasser and de Chadarevian, 2011, pp. 325–6).

4. In a 1959 letter, an MRC Officer considered that biological research would be 'even more different' than it already was in Cambridge after Sanger's

incorporation to the planned LMB (letter of R. Cohen to Max Perutz, The National Archives of the UK, file FD 7/1040).

5. The areas within which the prizes were awarded are different: Watson and Crick received the Nobel Prize in Medicine, while Perutz and Kendrew were awarded the Nobel Prize in Chemistry. Wilkins, at King's College London, shared the Prize with Watson and Crick, whereas Franklin did not receive the award due to her premature death from cancer and the Nobel being a distinction awarded only to active scientists.

6. Smith, a method-oriented scientist, had an unassuming personality, similar to Sanger's. Both were pioneers in the application of paper chromatography to nucleic acids (Bretscher and Travers, 2003).

7. Sanger, 1988. See also id., interview with author, Sanger Institute, Cambridgeshire, UK, 2005. Sanger's attitude towards professional identity contrasts with other contemporary scientists and is a key factor in the redefinition of his sequencing techniques. On the one hand, he was not as hostile to molecular biology as some other biochemists (e.g. Chargaff, 1978). On the other, he remained indifferent towards the expansion of the new discipline, unlike more enthusiastic colleagues such as Severo Ochoa (Santesmases, 2002). On molecular biology, biochemistry and professional identities see Abir-Am, 1992 and de Chadarevian and Gaudilliere, 1996.

8. The problem of the pervasiveness of retrospective accounts and of the view of a revolution is not exclusive to the historiography of molecular biology. Historians of computing have argued that the autobiographies of computer scientists, and the application to history of Moore's Law, have led to technological determinism and a notion of inevitability when discussing the progress of computerisation (Ceruzzi, 2005; Agar, 2003). Equally, Pnina Abir-Am and Clark Elliot have raised similar concerns about the effects of anniversaries or commemorations of different scientific disciplines, including molecular biology (Abir-Am and Elliot, 1999, Abir-Am 1999).

9. The circulation of scientists, instruments and theories across disciplinary boundaries and research environments is a recurrent concern of the history and philosophy of biology and science more generally. For a recent compilation on travelling facts see Howlett and Morgan, 2010 and especially Leonelli, 2010a; see also Secord, 2004.

10. Sanger's Laboratory Notebooks, Wellcome Trust Archives, London, Insulin Book 12, pp. 2377ff., Insulin Book 13; RNA books, files number SA/BIO/P/1/20ff. The interest in protein synthesis remained after Sanger's shift to DNA sequence determination in the 1970s.

11. The way in which Crick formulated the central dogma suggested that, from the structure of DNA, it would be possible to deduce the activity of proteins and, therefore, the functional aspects of the cell. Philosopher of biology, Sahotra Sarkar, has named this expectation 'reductionism' and shown how it has shaped the history of both genetics and molecular biology throughout the twentieth century (Sarkar, 1998, 1996b; Tauber and Sarkar, 1992). There are similarities between this sort of reductionism and the hypothesis of a direct connection between protein structure and function in Sanger's insulin work (see Chapter 1).

12. These efforts, as well as the development of RNA sequence determination, have been investigated in detail by historian Jerôme Pierrel. Apart from the

research in Cambridge, Pierrel has analysed the sequence determination work of Robert Holley at Cornell University, which was published before Sanger's. He has also explored Walter Fiers' 1976 determination of the RNA bacteriophage virus MS2, the first completed sequence of an organism, at the University of Ghent (Pierrel, 2009, 2012; Holley et al., 1965; Fiers et al., 1976).

13. Restriction enzymes to specifically cleave DNA generalised in the mid-1970s and became a key instrument in producing recombinant DNA molecules. Sanger used them to fractionate large DNA molecules in smaller fragments, but maintained the copying approach as the basis to determine the sequence of the fragments (see below).

14. Ways of teaching sequence determination techniques, plus the problems encountered in their circulation due to the specialised skills they required, are potential avenues for further research. Historian Sage Ross is addressing the difficulties in the spread of protein sequence determination in an ongoing PhD dissertation at the University of Yale.

15. Fred Sanger, interview with author, Sanger Institute, Cambridgeshire, UK, 2005.

16. Sanger's Laboratory Notebooks, Wellcome Trust Archives, London, Insulin, RNA and DNA Books (e.g. Insulin Book 10, p. 2140, RNA Books, file number SA/BIO/P/1/21, experiment R45, DNA Books, file number SA/BIO/P/1/42, experiment number D80). Sanger's assistants have also stressed the importance of these failed attempts and the 'determination' of their boss to develop suitable techniques (Alan Coulson, interview with author, LMB, Cambridge, UK, 2005).

17. Id., DNA Books, files number SA/BIO/P/1/40 to SA/BIO/P/1/41.

18. Ibid., file number SA/BIO/P/1/43, experiments number D80ff. Also see files number SA/BIO/P/1/31 to SA/BIO/P/1/34.

19. Holmes has applied this concept to the careers of Matthew Meselson, Franklin Stahl and Seymour Benzer, other key figures in the early development of molecular biology. In order to explore how scientific trajectories evolve on a day-to-day basis, he conducted interviews and analysed notebooks and published papers in detail. Holmes shares with this book the aim of determining the motivations behind scientific choices and of challenging retrospective accounts which reduce them to Eureka moments (Holmes, 2001, esp. pp. 3ff., 2006, esp.ch. 6, pp. 220–1). On history and the collective memory of scientists more generally see Abir-Am and Elliot, 1999.

20. The spread and ramifications of protein, RNA and DNA sequence determination show that the study of this form of work requires focusing on all historical attempts, rather than on retrospectively big names, such as Sanger. This book will contribute to such a collective historiography, but there is certainly more research to be conducted on sequence determination attempts contemporary to or preceding Sanger's.

21. The polymerase used by Sanger's group was, nevertheless, chemically modified, in order to make it suitable for the copying procedure (Coulson, personal communication, 2005). On Wu's DNA sequence determination efforts see Onaga, 2005. Kaiser's work was framed in a long-term investigation on λ which, in the early 1970s, led to collaboration with Paul Berg and to a series

of experiments which contributed to the technology of recombinant DNA. Historian Dogaab Yi has investigated Berg and Kaiser's collaboration with the aim of constructing a long history of experimental practices behind recombinant DNA, a similar goal to that sought for sequence determination in this book (Yi, 2008).

22. Historian Howard Chiang has shown the different identities that electrophoresis was given from its invention in the 1930s up to the 1960s, when it began being widely used by molecular biologists. The technique was originally invented as an instrument creating a 'moving boundary' in which the separation of the molecules was not complete. During the 1940s and 1950s, researchers in medical biochemistry and protein chemistry tested different separation media – filter paper, starch grain and agar gel – and created the concept of 'zone electrophoresis', in which the separation of the molecules was differentiated in independent areas on the surface. Electrophoresis, Chiang concluded, acquired its current identity as a 'molecular-sieving' technique in the early 1960s, when polyacrylamide gel emerged as the preferred separation medium (Chiang, 2009).

23. *Sequencing* was also used by Wu in one of his sticky ends articles in 1971 (Wu and Taylor, 1971, p. 491), but not systematically incorporated into the titles of published literature until Sanger's 1975 and 1977 papers. The gap between the use of this term in Sanger's laboratory notebooks and papers – which also occurred with sequence determination during the completion of insulin – suggests that Sanger rethought the terminology before publishing his experiments, deciding to name his work homogeneously as sequence determination in the mid-1950s and sequencing in the mid-1970s.

24. Brenner had already cooperated with Jacob in the discovery of messenger RNA, an achievement related to investigations of the coding problem during the late 1950s and early 1960s (Brenner, Jacob and Meselson, 1961). In 1961, biochemists Marshall Nirenberg and Heinrich Matthaei resolved the code's mechanism, matching a particular RNA sequence with a sequence of amino acids. This achievement was based on the synthesis of proteins by messenger RNA in the test tube and did not incorporate sequencing techniques (Kay, 2000, chs 5 and 6; Matthaei and Nirenberg, 1961; Nirenberg and Matthaei, 1961).

25. The differences in approach between biology and chemistry have also been noted by Kornberg, a Nobel Prize winner for his isolation of polymerase. Kornberg has defined chemistry and biology as 'two cultures', evoking the gulf that C.P. Snow postulated between the humanities and natural sciences (Kornberg, 1987, p. 6888; García-Sancho, 2008, pp. 198ff.).

26. Historian E.M. Tansey has shown how post-World War II biomedical research groups gradually expanded, with technicians and laboratory assistants becoming increasingly influential in research projects. Their roles are, nevertheless, difficult to trace, since they tend not to have left any published or archival record. Tansey has uncovered the work of a number of technicians and assistants based in the UK National Institute for Medical Research at different periods, during the second half of the twentieth century (Tansey, 2008). She has conducted oral histories, a main methodological tool employed in this book.

27. Sanger's Laboratory Notebooks, Wellcome Trust Archives, London, DNA Books, file number SA/BIO/P/1/43, quote from experiment D93.
28. A. Coulson, interview with author, LMB, Cambridge, UK, 2005.
29. Fred Sanger, interview with author, Sanger Institute, Cambridgeshire, UK, 2005.
30. Sanger's immersion in molecular biology was common among biochemists between the 1960s and 1970s. Historian María Jesús Santesmases has shown the incorporation of 'genetic thinking' into Severo Ochoa's work as he shifted his research agenda from enzymology to the coding problem in the early 1960s (Santesmases, 2002, pp. 193–4). Other researchers, however, preferred to maintain their biochemical identity.
31. This limitation of Sanger's technique was partially solved in the late 1970s, through the cloning of the template DNA (normally double-stranded) in a single-stranded bacteriophage called M13. The DNA to be sequenced was introduced into M13, and with the division of the virus, single-stranded templates were obtained (Sanger, 1988, pp. 23–4; id., 1980, pp. 437–9). The dideoxy method could not be directly applied to double-stranded DNA until the mid-1980s, with the advent of denaturation techniques to separate the strands.
32. F. Sanger and A. Coulson, interviews with author, Sanger Institute, Cambridgeshire, UK and LMB, Cambridge, UK, 2005.
33. This illustrates another extra-technological factor in the development of sequencing: the specifics of DNA. The ability of this molecule to express and duplicate itself meant that molecular biologists, and then Sanger, incorporated these mechanisms, considering them the natural and elegant way of sequencing and researching DNA. Proteins and RNA lack this capacity of self-replication and therefore were investigated through other procedures, such as chemical degradation.

Part II   Mechanisation – 1: Computing and the Automation of Sequence Reconstruction (1962–1987)

1. Crick's interest in protein sequence determination since his arrival at Cambridge suggests he might have taken the term sequence from Sanger, as formulated in the 1953 paper. There is no direct evidence for this, but it is likely that Crick read the term – associated with protein rather than DNA molecules – in Sanger's insulin papers or heard it in lectures, such as those he attended shortly after the formulation of the double helix (see Figure 2.1).
2. In a previous paper on the history of the concept of genetic information, I have likened this transition to a shift from the idea of a message to the idea of a text (García-Sancho, 2007a, pp. 21ff.). Doogab Yi, building on H.J. Rheinberger's scholarship, has suggested that the advent of recombinant DNA technologies in the mid-1970s – the same time sequencing was emerging – 'recast genes as entities that could not only be mobile but also reassembled, a sequence that could be rewritten thorough biochemical operations' (Yi, 2008, p. 613; see also Rheinberger, 1995).
3. Strasser has recently expanded his scholarship and explored the development of Genbank, the centralised DNA sequence database in the US. This

repository was established in 1982, two years after the foundation of its European counterpart (the EMBL Nucleotide Sequence Data Library, which I will address in Chapter 4). Strasser's investigation focuses on how computer analysis and circulation of sequence data affected long-standing scientific concerns such as credit, access and ownership (Strasser, 2011). My focus will be on the role of computing and databasing practices in the history of sequencing.

3   Sequencing Software and the Shift in the Practice of Computation

1. This argument is reinforced by Sanger's attitude once computers had been installed in his laboratory. According to his colleagues, Sanger tended to work on the computer 'very early in the morning', getting the computing tasks done 'before anyone else was around' (R. Staden, email interview with Glyn Moody, 2002).

2. The term *read-off* contrasts with concepts such as transcription or translation, coined over the 1950s and 1960s in the context of investigations on the coding problem (Kay, 2000, ch. 5). Reading off autoradiographs squares with a concept of information as sequence data, whereas transcription and translation are better adapted to a notion of information transfer from DNA to proteins.

3. On the notion of a genetic programme, its effects on *C. elegans* research and its consequences for the history of molecular biology see de Chadarevian, 1998b, 2000; García-Sancho, forthcoming. On Brenner's investigations on *C. elegans* more generally see Ankeny, 2000, 2001.

4. Computer-assisted bibliographic searches in indices and databases were also explored by Kendrew, who was involved in operations research due to his mobilisation in World War II, prior to his arrival to the Cavendish Laboratory (de Chadarevian, 2002, p. 119). For more details on computing, operations research and early biomedical uses see November, 2006, chs 1–2 and id., in press.

5. Strasser has placed the origins of these practices of 'collecting, comparing and computing' protein sequences in the 1950s, in the field of comparative biochemistry and the theoretical attempts at solving the coding problem. Comparative biochemists cross-examined protein sequences of various species with the aim of deducing their functional attributes. Ledley, together with Crick, was a member of the Tie Club, an informal group established by physicist George Gamow to decipher the genetic code. The group's members compared hypothetical and published protein sequences – sometimes with the help of computers – to test different theoretical models of the code (Strasser, 2010).

6. Computer-assisted phylogenetic trees were also constructed by population geneticists Anthony Edwards and Luigi Luca Cavalli-Sforza, but based on data on human blood group variation rather than protein sequences (Suárez-Díaz, 2010, note 11; Hagen, 2001). Comparative and immunological analysis of blood was another widespread tool for making evolutionary inferences that triggered controversies between different branches of evolutionary biology (id., 2010).

7. Needleman and Wunsch (1970) introduced dynamic programming into sequence comparison (Suárez-Díaz, 2010, pp. 79ff.). This programming strategy, used in operations research, consisted of deconstructing a complex task into partially overlapping subroutines. The programmer then devised an algorithm which minimised the number of necessary subroutines to fulfil the task. Dynamic programming presented conceptual similarities to sequence determination, as a practice engaged in reassembling and reducing the overlaps between certain subunits – protein fragments in the case of sequence determination; repetitive computing tasks in dynamic programming (García-Sancho, 2008, pp. 127ff.).

8. The minicomputer in which the *C. elegans* work was conducted was a Modular One commercialised by the British company Computer Technology Limited and acquired with funds from the UK Medical Research Council (MRC) at Brenner's request. This apparatus, and the attached graphics terminal, was pooled by the LMB Structural Studies Division to model the three-dimensional structure of proteins after calculation of their electron density (Correspondence between S. Brenner and MRC Officers J. Faulkner and J. Neale, 20th October 1969 to 4th March 1970, Archives of the LMB, Cambridge, UK, uncatalogued file on *Computers*). On the use of graphics terminals and image-processing programs by crystallographers, physiologists and protein chemists during the 1970s and 1980s see Francoeur and Segal, 2004; November, 2006, chs 4–5; Wieber, 2006.

9. Strasser has shown how evolutionary considerations were a means for the goals of comparative biochemists: they postulated that natural selection had left the sequences of the functionally relevant fragments of the proteins they compared unchanged. Molecular evolutionists, by contrast, used sequence variations to draw evolutionary conclusions about proteins (Strasser, 2010, p. 629). Hartley did explore the evolutionary implications of protein sequence comparisons in some of his papers (e.g. Hartley et al., 1965), but the main goal of his work was to relate sequences of the disulphide bridges of enzymes to their catalytic activity.

10. Levitt has retrospectively considered his rejection of Barrell's offer 'the greatest misjudgement' of his career. In his account, written in 2001, he acknowledges sequence analysis as a 'key part of modern computational biology' and quotes Needleman and Wunsch, as well as Sanger's work on ØX-174 (Levitt, 2001, p. 393). However, he does not refer to McLachlan.

11. R. Staden, email interview with Glyn Moody, 2002.

12. R. Staden, email interview with Glyn Moody, 2002.

13. R. Staden, email interview with Glyn Moody, 2002.

14. One of these researchers was John Walker who, following Hartley's move to Imperial College in 1974, had gained increasing responsibilities in the development of protein techniques within Sanger's division. Walker advised Staden in the 1979 version of his programs (Staden, 1977, p. 2609) and may have put him in contact with McLachlan.

15. This new program was run in the next generation of minicomputers commercialised by Digital Equipment Corporation, the VAX-11/780. The new apparatus had been purchased by the MRC in 1981 at Brenner's request and used for the *C. elegans* project and other lines of research at the LMB. The previous PDP-11 minicomputer – to which Staden adapted his early

programs – was installed at the LMB in 1975, originally to be connected to the graphics terminal used by both crystallographers and *C. elegans* researchers (S. Brenner, 1985: 'Future in-house computing at LMB'; Anonymous, 1982: 'Computer graphics at LMB' both in Archives of the LMB, Cambridge, UK, uncatalogued file on *Computers*; see also García-Sancho, 2012).

16. Communities of practice are becoming an object of increasing interest for STS scholars as a way of overcoming disciplinary barriers in the social studies of science and technology. A key context in which these communities are being investigated is the standardisation of data and its use in databanks (e.g. Ruhleder, 1995; Almklov, 2008; Zimmerman, 2008).

4   Sequence Databases and the Emergence of 'Information Engineers'

1. Olga Kennard, interview with author, Cambridge, UK, 2007. In a 1981 letter to an officer of the Medical Research Council (MRC, the UK Government body that funded the LMB), the then Director of the Laboratory, Sydney Brenner, considered that the DNA database was 'exactly the sort of thing that EMBL should be doing'. He advised the MRC officer to 'stand firm on doing it only through EMBL'. The reason for this was that British molecular biologists 'should have the use of it already paid for' by the contribution that the UK, as a Member State, provided for the operation of the EMBL. The MRC needed to make sure 'that we do not pay again for it' (The National Archives of the UK, Kew, document reference FD 7/2015).

2. The EMBL database has been the subject of a previous paper in which I explored the consequences of the newly created DNA sequence banks for the notion of *genetic information* (García-Sancho, 2011b). Here I will address the database algorithms, their designers and how they transformed the form of work characteristic of sequencing.

3. Various Authors (1975) 'European Molecular Biology Laboratory: Scientific Program and Indicative Scheme' in Papers and Correspondence of Sir John Kendrew, Bodleian Library, Oxford, UK, reference NCUACS 11.4.89/H263, quote from p. 11.

4. Quote from Various Authors (1978) 'EMBL Scientific Program and Indicative Scheme, 1980–83', p. 4; see also correspondence between J. Kendrew and K. Murray in Papers and Correspondence of Sir John Kendrew, Bodleian Library, Oxford, UK, references NCUACS 11.4.89/H264, H161 and H162. The researchers most involved in recombinant DNA and sequencing at the EMBL during the late 1970s were Hans Lehrach and Riccardo Cortese. The former had arrived from Harvard and established a line of research on the genetic expression of collagen and its role in the development of mice. Cortese moved to the EMBL from Sanger's laboratory in 1980 and devoted his research to making the dideoxy method compatible with double-stranded DNA, in order to correlate structure and function in transfer RNA genes. Protein sequence determination was also carried out at the EMBL, but by the late 1970s only one researcher was involved and the programme gradually lost support (European Molecular Biology Laboratory, 1981, pp. 40–2 and 79).

5. Graham Cameron's personal archive, European Bioinformatics Institute, Hinxton, Cambridgeshire, UK, folder on Schönau meetings. The debate on a large-scale sequencing initiative was a consequence of *Project K*, presented in the late 1960s by Crick as a potential research avenue for the future EMBL. The project sought the 'complete solution' of *E. coli* and proposed a full biological analysis of this organism including, among other potentially applicable techniques, DNA sequencing (Crick, 1973). Some European scientists expressed their reservations towards this initiative (Smith, 1974).

6. K. Murray, interview with author, School of Biological Sciences, University of Edinburgh, UK, 2007. Researchers in evolutionary biology expressed similar concerns about automation in the face of the introduction of computers into their field during the late 1960s and 1970s. At that time, taxonomists and molecular evolutionists maintained intense debates partly founded on whether evolutionary analyses should depend on the supposedly objective estimations of a computer (Hagen, 2001).

7. One of the researchers in charge of these local databases, Heinz Schaller at the University of Heidelberg, was the first to propose to Murray a centralised European DNA sequence repository (K. Murray, interview with author, School of Biological Sciences, University of Edinburgh, UK, 2007). During a second workshop in 1981, also held in Schönau, it was proposed that the heads of some local collections – Richard Grantham and Kurt Stüber at the Universities of Lyon and Cologne – should transfer their data to the EMBL sequence bank, the design of which was underway. In some cases, this transfer was slow and problematic (R. Grantham, undated: 'EMBL workshop on computing and nucleic acid sequences, 27–28th April 1981 at Schönau', in Graham Cameron's personal archive, European Bioinformatics Institute, Hinxton, Cambridgeshire, UK, folder on Schönau meetings).

8. K. Murray (1980) Conclusion letter to participants in the first Schönau workshop. In Graham Cameron's personal archive, European Bioinformatics Institute, Hinxton, Cambridgeshire, UK, folder on Schönau meetings.

9. K. Murray (1980) 'Vacancy notice' in Graham Cameron's personal archive, European Bioinformatics Institute, Hinxton, Cambridgeshire, UK, folder on Schönau meetings.

10. G. Hamm, phone interview with author, 2007. Before his appointment as database manager, Hamm was working with Lehrach's group, one of the pioneer teams in the introduction of DNA sequencing into the EMBL, where he optimised Gingeras and Staden's software (G. Hamm, 1980: 'Computer tools for nucleic acid sequence analysis'. In Papers and Correspondence of Sir John Kendrew, Bodleian Library, Oxford, UK, file number NCUACS 11.4.89/H.180).

11. Graham Cameron, interview with author, European Bioinformatics Institute, Hinxton, Cambridgeshire, UK, 2007.

12. Greg Hamm, phone interview with author, 2007.

13. The use of the concept of system in biology precedes World War II. Arantza Etxeberria and Jon Umerez have shown how Ludwig von Bertalanffy, founder of the general system theory in the social sciences, interacted during the 1930s with the Biotheoretical Gathering, an informal group of biologists based at the University of Cambridge. Bertalanffy himself worked on theoretical biology and wrote a book on the subject in 1928, translated into

English by the Biotheoretical Gathering member, J.H. Woodger. Theoretical biology as a field crystallised in the late 1950s and constructed a significant part of its identity as an alternative to molecular biology, being based on a holistic approach to the organism rather than a reductionist focus on DNA (Etxeberria and Umerez, 2006; see also Abir-Am, 1987; Haraway, 1976).

14. This transition triggered increasing qualifications to the central dogma of molecular biology, according to which information always flows one-directionally from DNA to RNA and proteins (see Chapter 2 and Antonarakis and Fantini, 2006). The last third of the twentieth century also witnessed a more general shift in the notion of computation – from linear processing of logical operations to network-multitask models – which affected other biological fields, particularly the brain sciences (Olazaran, 1996, pp. 643–8; Beaulieu, 2004; Stadler, 2010).

15. This new range of practices also epitomised a shift in the concept of genetic information from a pervasive metaphor which inspired and shaped biological research (e.g. Kay, 2000; Brandt, 2005; Rheinberger, 2006) to a research object or aim embodied in software tools systematically applied by professional systems scientists (García-Sancho, 2011b, pp. 81ff.). This shift reinforced the previously discussed transition of genetic information from an encoded message to a sequence of DNA which could be tackled with Sanger and Gilbert's techniques (see Introduction of Part II and García-Sancho, 2007a, pp. 24ff.).

16. Non-computerised biological repositories were common in natural history from the mid-sixteenth century onwards (Rosenberg, 2003; Secord and Jardine, 1996; Müller-Wille, 2003). During the late-nineteenth and early-twentieth century, epidemiologists and classical geneticists used mechanical and electrical devices to make calculations and process information. In 1966, one year after Dayhoff and Kennard's efforts, human geneticist Victor McKusick designed a computer-based repository of hereditary diseases (McKusick, 1966; Lindee, 2005, ch. 3).

17. Minicomputers and microcomputers coexisted with larger apparatus at the EMBL during the late 1970s and early 1980s. At that time, the Laboratory was connected to a 'large computer' at the Max Planck Institute for Nuclear Physics – also based in Heidelberg – used for complex calculations of molecular structures. Members of its Computing Group were considering the purchase of a 'midi' – medium size computer – which would provide both services to researchers and improved computational resources (Anonymous, 1979: 'Minutes of the First Meeting of the EMBL Computational Resources Committee, on 23rd and 24th March' in Papers and Correspondence of Sir John Kendrew, Bodleian Library, Oxford, UK, reference NCUACS 11.4.89/ H.198).

18. The increasing importance of data compilation in biomedicine has led Strasser to challenge this opposition, founded on the belief of a complete experimental shift in twentieth-century life sciences. Strasser, using the terms after Pickstone, proposes a 'hybridisation' in which natural historical and experimental ways of knowing are indistinguishable in current biomedical practice (Strasser, 2006, 2011, pp. 93ff.). Pickstone has noted the pervasiveness of analytical 'working knowledges' in current biomedicine and raised a possible 'de-differentiation' of disciplines around practices

of data analysis, half way between natural history and experimentation (Pickstone, 2007, pp. 513–4). The practice of collecting has been central to the history of the life sciences and molecular biologists benefited from medical collections in the mid-twentieth century (de Chadarevian, 1998; Strasser and de Chadarevian, 2011, pp. 325–6). However, from the 1960s onwards there has been an increasing shift from collecting specimens to the computer-assisted collection of data (García-Sancho, 2007a,b, 2008, pp. 236ff.). On the practice of collecting, its current scientific and financial importance, and its changing meaning in the history of the life sciences see Parry, 2004; Harvey and McMeekin, 2007.

19. Dayhoff's defeat, together with the funding difficulties she and Kennard experienced, may be analysed from a gender perspective (Strasser, 2010, pp. 649ff.). The routine activities of compiling and organising data in administration, business and science had traditionally been conducted by women, who became important actors in the early stages of computerisation (Light, 1999; de Chadarevian, 2002, pp. 118–27). However, professional communities and funding agencies within biology were – and largely still are – dominated by men. The prominence of women among the early users of computers did not result in a significant female presence when the importance of computerised data analysis in biology increased and the first DNA sequence databases emerged during the early 1980s.

20. The relational model was invented during the 1960s by IBM programmer E.F. Codd, and refined and popularised in the subsequent decades by C.J. Date (Codd, 1970; Date, 1981 [1975], 1990 [1987]). It was the basis for Oracle, a successful database multinational created in 1977 in Silicon Valley and founded by Larry Ellison, a self-taught, non-disciplinary entrepreneur similar to Hamm and Cameron (Wilson, 1997). During the late 1980s, the EMBL reached an agreement with Oracle for assistance in the management of its database. The relational model was then implemented in the Nucleotide Sequence Data Library.

21. G. Hamm, phone interview with author, 2007; G. Cameron, interview with author, European Bioinformatics Institute, Hinxton, Cambridgeshire, UK, 2007.

22. Stüber's database was promoted by, among others, P.H. Schreier, who had worked with Cortese at Sanger's laboratory and then migrated to a group engaged with DNA sequencing in Cologne (Schreier and Cortese, 1979). Whereas Cortese moved to the EMBL in the early 1980s, Schreier joined another node of sequencing development at the University of Cologne.

23. G. Hamm, phone interview with author, 2007 and id. (1980) 'Computer tools for nucleic acid sequence analysis' in Papers and Correspondence of Sir John Kendrew, Bodleian Library, Oxford, UK, file number NCUACS 11.4.89/ H.180.

24. Despite computer editors and other related applications being available from the 1960s, Haigh has shown that the processing of texts was initially confined to typewriters, teletypes and dictating machines. The main use of computers in the office at that time was to correlate discrete data such as figures – prices or stocks – and qualitative labels – names of employees or products. Text processing in centralised mainframes or shared minicomputers was considered too expensive, since it consumed much valuable computer time. With the

emergence of the personal computer in the mid-1970s, word processing programs became widespread and the manipulation of text gradually became the main computing application (Haigh, 2006a; Bergin, 2006a,b).

25. G. Cameron (1984) 'Procedure for polishing. EMBL Nucleotide Sequence Data Library', in id. personal archive, European Bioinformatics Institute, Hinxton, Cambridgeshire, UK, folder on memos and reports.

26. The association of interrelated databases in higher order collections – for example DNA and protein sequence databases in a gene expression bank and software package – has made bio-ontologies an essential instrument in current biology, and a main subject of STS research. Bio-ontologies are tools that standardise the terminology and format of entries from different repositories, easing connections between them (see Leonelli, 2008, 2010b).

27. G. Cameron, interview with author, European Bioinformatics Institute, Hinxton, Cambridgeshire, UK, 2007. Some journal editors with whom the EMBL team attempted to negotiate sequence submission to the database prior to publication, raised reservations about this proposal. Robert Cook-Deegan has shown that as late as 1988, the Editor of *Nature* resisted mandatory database submission of the sequences to be published in his journal. The reasons he gave – scarce computer facilities in many laboratories, that the sequences had 'nothing to do with the content' of *Nature's* papers and that the journal did not need to depend on a database – point to a disregard of the Data Library staff, similar to that the EMBL biologists were beginning to overcome at that time (Cook-Deegan, 1994, pp. 288–9). These reservations were also due to issues regarding the attribution of credit, ownership of the sequences, and the limits of access and distribution. The resolution of these problems, as Strasser has shown, required long and complex negotiations (Strasser, 2011, pp. 81–90).

28. This situation was not so crude in the LMB, where the software community had a more academic and biological background. During the mid-1980s, Staden associated with other LMB biologists who were investigating the nematode worm *C. elegans* and required computer expertise (see Chapter 5). The worm team was well respected among the community of molecular biologists and Staden, together with Gingeras and McLachlan, has been retrospectively considered a *pioneer* of bioinformatics (e.g. Roberts, 2000).

29. The growing sophistication, changing definitions and pervasiveness of the concept of the gene have been recurrent topics in the philosophy and history of biology (Beurton, Falk and Rheinberger, 2000; Fox Keller, 2000; Moss, 2004; Griffiths and Stotz, 2006; López-Beltrán, 2004; Müller-Wille and Rheinberger, 2007; Taylor, 2008). An increasing number of STS scholars have also addressed the problem of 'producing meaning' from the information stored in biological databases (quote from Fujimura 1999; see also Fujimura and Fortun 1996; Fortun 2008; Leonelli 2008, 2010a,b).

30. G. Hamm and G. Cameron (1982 to 1985) 'Nucleotide Sequence Data Library Staff', 'Report for the review of the Computing and Applied Mathematics Programme', 'Minutes of the 2nd meeting. Nucleotide Sequence Data Library. 26th March 1984', 'Draft Development Plan' and quotes from 'Biological expertise for the Data Library', all in Graham Cameron's personal archive, European Bioinformatics Institute, Hinxton, Cambridgeshire, UK, folder on reports and memos. In one section of

'Biological expertise for the Data Library', a report written in 1985, the task of reviewing an entry was compared with that of 'refereeing a short paper'. This suggests that a meta-literature of entries was arising among the database staff, with cross-referencing and a quality control similar to those of scientific publications.

31. G. Hamm (1985) 'Biological expertise for the Data Library' in Graham Cameron's personal archive, European Bioinformatics Institute, Hinxton, Cambridgeshire, UK, folder on reports and memos. The insufficiency of the EMBL computing resources for providing the desired service role to European molecular biologists was not a novelty in the mid-1980s. From 1979 onwards, a Computer Users Group and a Computational Resources Committee had been established to advise the Laboratory's computing staff on the best ways to fulfil biological requests. In a letter written to the Chairman of the Committee, Sydney Brenner, during preparations for the first meetings, an EMBL representative admitted that the situation at the Laboratory's Computing Group was 'not healthy': the staff was primarily engaged in the development of computational technologies and job vacancies for personnel specifically devoted to service tasks were not filled, due to a lack of applicants. This had created 'friction' between the computing staff and biologists, as well as frustration for the latter (letter of R.C. Ladner to S. Brenner, 7th March 1979. In Papers and Correspondence of Sir John Kendrew, Bodleian Library, Oxford, UK, reference NCUACS 11.4.89/ H.198).

32. Various Authors (1982) 'Scientific Program for EMBL, 1983–1987' in Papers and Correspondence of Sir John Kendrew, Bodleian Library, Oxford, UK, reference NCUACS 11.4.89/H.183, quotes from p. 6 and pp. 13–14.

33. Ibid. and 'Summary Report of Meetings of the Senior Scientists' (1981–82) in Papers and Correspondence of Sir John Kendrew, Bodleian Library, Oxford, UK, reference NCUACS 11.4.89/H.183. Murray was among the EMBL Senior Scientists who participated in these meetings, which were chaired by Philipson. As the final supervisor of the Data Library, Murray reported on the database team's progress.

34. Anonymous (1986) 'Report on the workshop on biocomputing at the EMBL. Heidelberg, November 21st-22nd 1985' in The National Archives of the UK, Kew, document reference FD 7/2015.

35. On the absorption of preceding practices by emerging biological disciplines see Powell et al., 2007 and Strasser, 2010.

5 A New Approach to Sequencing at Caltech

1. See also W. Dreyer, interview with S.K. Cohen, Oral History Project, The Caltech Archives, Pasadena, US, 1999.

2. The mechanisation of protein sequence determination after Stein and Moore's automation of amino acid analysis (see Chapter 1) is a potential avenue for further research. I am indirectly addressing it in the course of my current investigations on the circulation and use of sequencing in different contexts, mainly Spanish biomedical research. In this book, however, I will focus on Caltech and ABI's DNA sequencer, which better suits the comparison of forms of work I aim to establish.

3. L. Smith (2008) 'The development of automated DNA sequencing', unpublished paper.

4. L. Hood 'Immunology, disease and Caltech' (1977); 'Genetic engineering and medicine of the future' (1982); 'Biotechnology in the future' (1983); 'Medicine in the 21st century' (1988). The Caltech Archives, Pasadena, US, Audio Tapes Collection.

5. Ibid. In further autobiographical accounts, Hood has maintained this view and described 'five paradigm changes' in biology partially fostered by his work: (1) the integration of biological research and instrumentation; (2) cross-disciplinary biology; (3) the Human Genome Project and discovery science; (4) systems biology, and (5) predictive, personalised, preventive and participatory (P4) medicine (Hood, 2002; quote from id. 2008, p. 1.1).

6. A. Coulson (undated) 'Chain terminator sequencing' in Papers and Correspondence of Alan Coulson, Archives of the MRC Laboratory of Molecular Biology, Cambridge, UK. Currently being catalogued at the Wellcome Trust Archives, London, UK. Provisional reference number PP/COU. Staden's early programs were equally made available as tapes 'on request' (Staden, 1977, p. 4037; 1979, p. 2601). On the role of protocols and their circulation see Lynch et al., 2008, pp. 87ff.

7. On the history of gene mapping see Gannett and Griesemer, 2004; Kass and Bonneuil, 2004; Gaudillière, 2004b; Lindee, 2005; Harper, 2008, ch. 7.

8. Papers and Correspondence of A. Coulson, Archives of the MRC Laboratory of Molecular Biology, Cambridge, UK. Currently being catalogued at the Wellcome Trust Archives, London, UK. Provisional reference number PP/COU.

9. Sulston and Coulson's expertise transformed the LMB into a 'centre of calculation' or 'obligatory point of passage', both for obtaining mapping information and for learning the techniques (Latour, 1987, ch. 6; 1988, pp. 59–60). The laboratories which requested mapping information worked on genetic anomalies in the worm's development and behaviour. This suggests that Sulston, Coulson and Waterston's efforts did not strictly change the goals of the *C. elegans* project, but rather the way of achieving them (García-Sancho, 2012). For a similar exchange network and rhetoric of free circulation of materials around the *Drosophila* fly in the early-twentieth century see Kohler, 1994.

10. Historian of technology David Edgerton has critically summarised these views as 'declinism', showing that the relationship between science and technology in Britain and the US during the twentieth century – and more generally between pure and applied research – is far more complex (Edgerton, 1996, 1987).

11. The 'scandals' around the patents of penicillin and monoclonal antibodies in Britain have been critically revisited by Jonathan Liebenau, Robert Bud and Soraya de Chadarevian (Liebenau, 1987; Bud, 1998a, 2009; de Chadarevian, 2002, pp. 353–62). A number of historians have conversely shown a long academic tradition in twentieth-century US science, particularly in biology (Reingold, 1981, 1991; Pauly, 2000; Benson, Maienschein and Rainger, 1988, 1991).

12. Caltech received 83 million dollars in public research contracts during the war. In the post-war and Cold War periods, the Jet Propulsion Laboratory

at Caltech became one of the main contractors of the US Administration and an active collaborator with Hood, Dreyer and Hunkapiller in the development of the automatic protein sequenator (Leslie, 1993, pp. 4 and 6–8). Engineers at the Jet Propulsion Laboratory showed Caltech biologists an application of mass spectrometry designed to detect life in space which could also be used to identify amino acids (Caltech, 1977–79; W. Dreyer, interview with S.K. Cohen, Oral History Project, The Caltech Archives, Pasadena, US, 1999).

13. The development of the DNA sequenator and the use of ABI's commercial sequencer by public laboratories during the HGP triggered a media controversy. In 2000, the US Government accused Hood's group of having used Federal funds in the development of the sequenator and urged ABI to return part of the public money invested in the purchase of DNA sequencers during the HGP. Hood and other Caltech representatives admitted having received a National Science Foundation grant in 1985 'to automate DNA sequencing', but claimed the money only arrived when the sequenator was already finished (Gosselin and Jacobs, 2000).

14. A scandal about alleged data fabrication by two members of Hood's laboratory at Caltech in 1990 led both administrators and some fellow biologists to question the governance of such a large and diversified group (Kumar et al., 1991). By that time, Gates had begun to invest in biotechnology start-ups and saw this field as complementary to the information technology market. He decided to commit 12 million dollars to the University of Washington after listening to three public lectures delivered by Hood in Seattle (Dietrich, 1992).

15. Various authors (1963) 'The laboratory of molecular biology: proposals for extension' in Francis Crick's Papers, Wellcome Trust Archives, London, UK, file number PP/CRI/B/1/1. There are signs, however, that the current proliferation of paperwork and its disruptive effect on research may lead some funding bodies back towards more open ended support for notable individuals, in line with the early LMB scheme.

16. On the LMB and Caltech as particular research 'cultures' see de Chadarevian, 2002, especially ch. 9, and Kay, 1993.

17. In a historical investigation of the German genetics community, Jonathan Harwood has proposed the concept of *styles of thought*. He distinguishes various research schools in early twentieth century Germany marked by particular 'ontological and/or epistemological assumptions', sometimes due to the schools' leaders and sometimes to institutional habits. As in the cases of Caltech and the LMB, these assumptions affect the organisation of the different institutes within each school, 'their sources of funding and their responses to political pressure' (Harwood, 1993, pp. 9–10 and 7). John Pickstone and Jean Paul Gaudillière further showed how these styles of thought may be interrelated to different 'ways of working' or 'styles of work' (Pickstone, 2000, 2007; Gaudillière, 2004a).

18. L. Smith (2008) 'The development of automated DNA sequencing', unpublished paper.

19. Radioactivity only became a problem in the context of the automation of sequencing and neither Sanger nor Gilbert considered it dangerous or inconvenient for their techniques. The different research cultures at Caltech

and the LMB, as well as the proliferation of anti-nuclear movements in the 1980s, may have played a role in these divergent attitudes.

20. Historical File on Leroy Hood, The Caltech Archives, Pasadena, US.

21. T. Hunkapiller, interview with author, Institute for Systems Biology, Seattle, US, 2006.

22. L. Smith, interview with author, Seattle, US, 2006.

23. Dreyer and other commentators maintain that fluorescent labelling was considered by Huang in the course of his 1980 automation attempt. He even ordered some dyes, but never incorporated them into a viable sequencing technique (Gosselin and Jacobs, 2000, p. A19; W. Dreyer, 2005, interview with S.K. Cohen, Oral History Project, The Caltech Archives, Pasadena, US, transcript pp. 89–90). Smith was originally hired by Hood with the aim of applying his advanced techniques of fluorescent labelling to antibodies. He subsequently shifted from immunological research to become increasingly involved with the sequenator (L. Smith, 2008: 'The development of automated DNA sequencing', unpublished paper).

24. T. Hunkapiller, interview with author, Institute for Systems Biology, Seattle, US; see also Cook-Deegan, 1994, p. 68 and Wills, 1991, p. 152.

25. The fact that these new strategies derived from the dideoxy method made Sanger's sequencing the basis for automation, unlike in the case of the protein sequenator, which had been based on Edman's technique (see Chapter 1). Nevertheless, Huang had based his previous attempt on Gilbert's sequencing, and the Hunkapiller brothers and Smith initially explored the possibilities of this technique.

26. L. Smith (2008) 'The development of automated DNA sequencing', unpublished paper.

27. Anonymous (1984) 'Review of all R&D activities' in ABI Archives, Foster City, US, Steven Fung's Collection.

28. K. Connell (1984) 'DNA sequencing' in ABI Archives, Foster City, US, Steven Fung's Collection.

29. This reciprocal influence suggests that the history of connections between biology and computing has operated two-directionally. Understanding such connections historically and two-directionally problematises the increasing STS literature on an alleged 'natural marriage' between biology and computing, or on the 'digital shaping' of biology (e.g. Cook-Deegan, 1994, pp. 285ff.; Lenoir, 1999; Moody, 2004; Zweiger, 2001). It also challenges the rigid and artificial separation between 'wet' and 'dry' biology (see Chow-White and García-Sancho, 2012; Penders et al., 2008; November, 2006).

30. The design of this apparatus was not prevented by the reservations towards the automation of sequencing expressed at the EMBL's Schönau workshops (see Chapter 4). However, the way in which the sequencing process was portrayed in those meetings – as always subordinated to the judgement of the user biologist – may have shaped the framing of Ansorge's sequenator in the *autoradiograph world*. During the negotiations of the EMBL Scientific Program for the years 1983 to 1987, the possibility was raised of including 'research projects of direct relevance to biotechnology'. Nevertheless, it was considered 'unwise to divert substantial part of the resources to entirely new programmes of applied research'. This ambivalence suggests that Ansorge's support was weaker than that received by Hood at Caltech,

despite the EMBL's historical connections with private industry easing the transformation of the prototype sequenator into a commodity to rival that of ABI during the late 1980s (quotes from Various Authors, 1982: 'Scientific Program for EMBL, 1983–1987' in Papers and Correspondence of Sir John Kendrew, Bodleian Library, Oxford, UK, reference NCUACS 11.4.89/H.183, p. 14).

31. The coexistence of manual and automatic sequencing as alternative forms of work reflects what David Noble has called different 'technological possibilities'. In analysing the automation of the US metal industry, he has argued against its inevitability and suggested it should be seen as the result of a series of choices shaped by a dominant capitalist ideology (Noble, 1984). The rise of the sequenator may also be seen as a choice of Caltech's approach to sequencing over those of their rivals, shaped by the triumph of Hood's group attitude towards automation.

6 The Commercialisation of the DNA Sequencer

1. This term 'university-industrial complex' was used by biologist Martin Kenney in a 1986 book about biotechnology. With it, he paraphrased the 'military-industrial complex' which US President Dwight Eisenhower had warned against in his 1961 farewell address (Kenney, 1986). Historian Jean-Paul Gaudillière has suggested the concept of 'way of regulating' – inspired by Pickstone's 'ways of knowing' and 'working' (Pickstone, 2001) – to better explain the emergence of biotechnology and its transition from molecular biology (Gaudillière, 2009, p. 20; see Introduction).

2. Biogen was focused on isolating insulin-producing genes. It used sequencing for this aim, but never attempted to develop commercial sequencers. This suggests that Gilbert – to some extent shaped by Harvard's institutional environment – shared the attitude towards automation of the LMB and saw sequencing as a tool applicable to human-led research rather than as a technique that could result in a commercial apparatus (see Chapter 5). Other leading figures in manual sequencing, such as Carl Weissmann and Kenneth Murray, were also involved in the early history of Biogen (Hofschneider and Murray, 2001).

3. S. Eletr (1981) 'Business summary' and 'GeneCo Charter' in ABI Archives, Foster City, US, André Marion's Collection.

4. The 1981 pre-foundational documents stated that the company would 'remain abreast of the field in order to supply its future needs for instrumentation and reagent systems' in 'clinical diagnostics, therapeutics and bio-manufacturing'. These areas were always regarded as hypothetical and secondary to the automatic sequencers and synthesisers, which were manufactured as soon as the company initiated its activity (S. Eletr, 1981, 'GeneCo Charter' in ABI Archives, Foster City, US, André Marion's Collection).

5. Robert Ledley, a pioneer in the introduction of computers into biology, had also worked on medical instrumentation and proposed the use of computers to automate diagnosis during the 1950s and 1960s (see Chapter 3). This, together with Eletr's role at HP, suggests that medicine played a key role in the confluence of electronic equipment and biological processes (see November, 2006, ch. 2).

6. A. Marion, interview with author, Applied Biosystems, Foster City, US, 2006.

7. Annonymous (undated) 'DNA sequencing R+D. Organisational chart' in ABI Archives, Foster City, US, André Marion's Collection. Quotes from A. Marion, interview with author, Applied Biosystems, Foster City, US, 2006.

8. Anonymous (1986) 'Training update' in ABI Archives, Foster City, US, Library Collection.

9. Kenney's 1986 account of biotechnology stated that the new start-ups were 'information intensive' and squared with a society increasingly dependent on information and personal computers (Kenney, 1986, p. 4). In a recent and still unpublished paper, Christophe Bonneuil and Jean-Paul Gaudillière show that current biomedical research increasingly tends towards organisational models based on 'team networks' rather than 'assembly lines'. These networks, they argue, are based on the architecture of the personal computer and their impact on the practice of biomedicine is equivalent to that of cybernetics on biology after World War II (Bonneuil and Gaudillière, 2007).

10. S. Fung, interview with author, Applied Biosystems, Foster City, US, 2006.

11. Anonymous (1987) 'Model 370A DNA sequencer preliminary users manual', pp. 2–3, in ABI's Archive, Foster City, US, Library Collection.

12. S. Fung, interview with author, Applied Biosystems, Foster City, US, 2006. On the importance and shaping power of representation in the history and philosophy of science see Lynch and Woolgar, 1988. On biological representation in particular, Suárez-Díaz, ed. 2007.

13. Various authors (1987 and 1988) '370A DNA sequencer agenda' and 'DNA sequencing – agenda' in ABI's Archives, Foster City, US, Steven Fung's Collection.

14. The gel view, paradoxically, did not correspond with a mimetic representation of the DNA fragments. Due to their variable mobility on the gel (see Chapter 5), Caltech and ABI's researchers had introduced 'correction factors' in the migration of the fragments (Smith et al., 1986, p. 678). The colour spots, consequently, appeared regularly aligned on the computer screen when this regularity did not occur in the DNA fragments after their separation on the gel surface.

15. Another PCR use licensed by ABI was forensics, which in the 1990s became a main niche of the technique and the subject of increasing STS investigations (Erlich, Gelfand and Sninski, 1991; Jordan and Lynch, 1993; Lynch et al., 2009; Gannett, 2004). On the merger of ABI with Perkin-Elmer see Various Authors (1992) 'The merger' in ABI Archives, Foster City, US, André Marion's Collection.

16. Legal battles over the financial rights of biotechnology applications were common in the US from the second half of the 1980s onwards. A widely publicised example was between Cetus and chemical multinational DuPont, which in the early 1990s challenged PCR's patent, claiming the technique had been in the public domain since the early 1970s, when molecular biologist H.G. Khorana published a journal paper describing a procedure based on enzymatic duplication of DNA. Cetus won the case (Mullis, 1994, 1998, ch. 3).

17. The introduction of robotics into DNA mapping and sequencing was pioneered by Généthon, created in 1991 as an off-shoot of the French Centre for the Study of Human Polymorphism (CEPH, in French). This Centre was

a private charity promoted by the French Muscular Dystrophy Association. During the second half of the 1980s, the CEPH had mapped the genes involved in this and other hereditary diseases, and in the following decade Généthon was launched as a platform for large-scale human DNA mapping. In 1996, a large-scale sequencing centre called Génoscope was created in France (Rabinow, 1999; Kaufmann, 2004; Ramillon, 2007, ch. 3).

18. The name *genomics* was suggested to McKusick and Ruddle by another geneticist, T.H. Roderick, at a meeting in 1986 (Kuska, 1998). *Genome* is an older term, coined in the 1920s by German botanist, Hans Winkler, as a word combining *gene* and *chromosome*. It became common in the late 1970s with the sequencing of bacteriophage viruses by Sanger's group, and in the 1980s, both geneticists and molecular biologists widely used genome to refer to the full DNA molecule of an organism (Lederberg and McCray, 2001). On the naming histories of genetics, molecular biology and genomics see Powell et al., 2007.

19. During the mid- to late 1960s, McKusick had been involved in *Mendelian Inheritance of Man,* a database of hereditary genetic diseases at Johns Hopkins Hospital (see Chapter 4). He also proposed locating the genes in human chromosomes using the new physical mapping techniques, and formed an international human genome mapping community with periodical meetings and exchanges of information. The role of human and medical geneticists such as McKusick and Ruddle in the configuration of genomics is a potential avenue for further research, which is being addressed by both geneticists and historians (e.g. Harper, 2008, ch. 7; Lindee, 2005; Bodmer and McKie, 1997).

20. Anonymous (1999) 'ABI Prism 377 DNA Sequencer: Greater Versatility no Matter how you Measure It' and 'ABI Prism 3700 DNA Analyser' in ABI Archives, Foster City, US, Library Collection.

21. Sociologist Andrew Barlett has explained the success of the sequencing initiative at the Sanger Centre as the 'recruitment of sentiment'. Workers at this institution engaged with its values, namely those regarding to the prior mapping of the genome and public availability of the sequence (Barlett, 2008, part three). This suggests that the organisation of labour at the Sanger Centre was shaped by a particular culture, in some ways different and in some ways connected with that of the LMB.

22. Fortun has partially challenged this dichotomy by showing that the international and supposedly *public* consortium also involved pharmaceutical companies. However, he has conceded that Celera had a more markedly commercial orientation than the consortium (Fortun, 1993, 2008, pp. 31ff.).

Conclusions: A Long History of Practices

1. Historians have soundly criticised the caricatured image of molecular biology – as a discipline engaged in a reductionist approach to DNA – which resulted from these retrospective accounts (e.g. de Chadarevian, 2002, ch. 8; Strasser, 2002). The accounts, however, can be fruitfully used as evidence to reconstruct the history of both molecular biology and genomics (Abir-Am, 1985; Suárez-Díaz, 2010, pp. 69ff.).

2. This conflation mirrors a similar one that philosopher of biology Lenny Moss has denounced between two different concepts of the gene. According to him, during the second half of the twentieth century, biologists increasingly superposed Gene-D – a developmental gene which corresponds with a nucleotide sequence – onto Gene-P – a preformationist gene implying a phenotypic effect. In these concepts, genes possess different properties and are expressed through different regulatory mechanisms. Therefore, they cannot, as many biomedical researchers appear to believe, be attributed to a single entity (Moss, 2004, ch. 1).

3. These bidirectional interactions have also been proposed by Joseph November and Soraya de Chadarevian. They have also demonstrated that medical diagnosis, physiology and genetics were the main fields for the early introduction of computers into the life sciences in the 1960s, rather than molecular biology (November, 2006; de Chadarevian, 2009, 2002, ch. 4).

4. STS scholars are now intensely investigating the regimes of hope around genomics, the public use of informational metaphors and the collaboration between biologists and computing experts (e.g. Brown, Kraft and Martin, 2006; Fortun, 2008; Bostanci, 2010; Leonelli, 2010a,b; Penders et al., 2008; Kaye, ed. 2012). Their scholarship highlights the importance of perceived pasts and previous trajectories in these issues, but has not yet systematically incorporated a historical perspective. The history of sequencing I have presented – or the study of other genomic techniques – may in the future be circumscribed to particular institutional settings and fruitfully combined with discourse analysis or anthropological work. This combination would show how historically constructed promises are managed, transformed and communicated among different communities, or how the divergent careers of biologists and computer experts shape their interactions in biocomputing centres.

5. For a case study on the introduction of sequencing as a central and shared facility see García-Sancho, 2011a. On the pooling of biomedical techniques in universities more generally see Wilson, 2008; Wilson and Lancelot, 2008 and Lancelot, 2007.

Bibliography

Abelson J. (1980) 'A revolution in biology' in *Science,* 209: 1319–21.

Abir-Am P. (1982) 'The discourse of physical power and biological knowledge in the 1930s: a reappraisal of the Rockefeller Foundation "policy" in molecular biology' in *Social Studies of Science,* 12: 341–82.

Abir-Am P. (1985) 'Themes, genres and orders of legitimation in the consolidation of new scientific disciplines: deconstructing the historiography of molecular biology' in *History of Science,* 23(1): 73–117.

Abir-Am P. (1987) 'The biotheoretical gathering, trans-disciplinary authority and the incipient legitimation of molecular biology in the 1930s: new perspective on the historical sociology of science' in *History of Science,* 25: 1–70.

Abir-Am P. (1991) 'Noblesse oblige: lives of molecular biologists' in *Isis,* 82(2): 326–43.

Abir-Am P. (1992) 'The politics of macromolecules: molecular biologists, biochemists and rhetoric' in *Osiris,* Second series, 7: 164–91.

Abir-Am P. (1999) 'The first American and French commemorations in molecular biology: from collective memory to comparative history' in *Osiris,* Second series, 14: 324–70.

Abir-Am P. and Elliot C.A. (eds, 1999) *Commemorative practices in science: historical perspectives on the politics of collective memory,* special issue of *Osiris,* Second series, 14.

Agar J. (2003) *The Government Machine: A Revolutionary History of the Computer* (Cambridge: MIT).

Allibone T.E. (1984) 'Metropolitan-Vickers Electrical Company and the Cavendish Laboratory' in J. Hendry (ed.) *Cambridge Physics in the Thirties* (Bristol: Adam Hilger): 150–73.

Almklov P.G. (2008) 'Standardised data and singular situations' in *Social Studies of Science,* 38(6): 873–97.

Anderson S., Bankier A.T., Barrell B., de Bruijn M.H.L., Coulson A., Drouin J., Eperon I.C., Nierlich D.P., Roe B.A., Sanger F., Schreier P.H., Smith A., Staden R. and Young I.G. (1981) 'Sequence and organisation of the human mitochondrial genome' in *Nature,* 290: 457–65.

Angell M. (2005) *The Truth about the Drug Companies: How They Deceive Us and What to Do about It* (New York: Random House).

Ankeny R. (2001) 'The natural history of *C. elegans* research' in *Nature Review Genetics,* 2: 474–9.

Ankeny R. (2010) 'Historiographical reflections on model organisms: or, how the mureaucracy may be limiting our understanding of contemporary genetics and genomics' in *History and Philosophy of the Life Sciences,* 32(1): 91–104.

Ansorge W., Sproat B., Stegemann J., Schwager C. and Zenke M. (1986) 'A nonradioactive automated method for DNA sequence determination' in *Journal of Biochemical and Biophysical Methods,* 13(6): 315–23.

Ansorge W., Sproat B., Stegemann J., Schwager C. and Zenke M. (1987) 'Automated DNA sequencing: ultrasensitive detection of fluorescent bands during electrophoresis' in *Nucleic Acids Research*, 15(11): 4593–602.

Antonarakis S. and Fantini B. (eds, 2006) *History of the central dogma of molecular biology and its epistemological status today*, special issue of *History and Philosophy of the Life Sciences,* 28(4).

Applied Biosystems (1983–92) *Annual Reports* (Foster City: Applied Biosystems).

Applied Biosystems (1990a) 'Applied Biosystems obtains licence for PCR technology' in *Biosystems Reporter,* 7: 1.

Applied Biosystems (1990b) 'New system for automated DNA analysis' in *Biosystems Reporter,* 8: 13.

Applied Biosystems (1990c) 'European symposia shows growing acceptance of automated DNA sequencing' in *Biosystems Reporter,* 8: 11.

Applied Biosystems (1990d) 'Three new DNA sequencing kits smooth road to automation' in *Biosystems Reporter,* 10: 10.

Armon R. (2012) 'Between biochemists and embryologists: the biochemical study of embryonic induction in the 1930s' in *Journal of the History of Biology*, 45: 65–108.

Astbury W. (1934) 'X-ray studies of protein structure' in *Cold Spring Harbor Symposia on Quantitative Biology*, 2: 15–27.

Atkinson P., Greenslade H. and Glasner P. (eds, 2007) *New Genetics, New Identities* (London and New York: Routledge).

Austoker J. and Bryder L. (eds, 1989) *Historical Perspectives on the Role of the MRC: Essays in the History of the Medical Research Council of the United Kingdom and Its Predecessor, the Medical Research Committee, 1913–1953* (Oxford: Oxford University Press).

Backus J. (1978) 'The history of Fortran I, II and III' in R.L. Wexelblat (ed.) *History of Programming Languages I* (New York: ACM): 25–74.

Bairoch A. (1991) 'The SWISS-PROT protein sequence data bank' in *Nucleic Acids Research,* Supplement 20: 2019–22.

Balmer B. (1996a) 'Managing mapping in the Human Genome Project' in *Social Studies of Science,* 26: 531–73.

Balmer B. (1996b) 'The political cartography of the Human Genome Project' in *Perspectives on Science,* 4(3): 249–82.

Balmer B. (1998) 'Transitional science and the Human Genome Mapping Project Resource Centre' in P. Glasner and H. Rothman (eds) *Genetic Imaginations: Ethical, Legal and Social Issues in Human Genome Research* (Aldershot: Ashgate): 7–19.

Balmer B. and Sharp M. (1993) 'The battle for biotechnology: scientific and technological paradigms and the management of biotechnology in Britain in the 1980s' in *Research Policy,* 22: 463–78.

Barnes B. and Dupré J. (2008) *Genomes and What to Make of Them* (Chicago: University of Chicago Press).

Barnett C. (1986) *The Audit of War: Illusion and Reality of Britain as a Great Nation* (London: Macmillan).

Barrell B. (1991) 'DNA sequencing: present limitations and prospects for the future' in *FASEB Journal,* 5: 40–5.

Bartlett A. (2008) *Accomplishing Sequencing the Human Genome.* PhD dissertation, ESRC Cesagen Centre for the Economic and Social Aspects of Genomics, University of Cardiff.

Beatty J. (2000) 'Origins of the US Human Genome Project: changing relationships between genetics and national security' in P.R. Sloan (ed.) *Controlling our Destinies: Historical, Philosophical, Ethical and Theological Perspectives on the Human Genome Project* (Indiana: University of Notre Dame): 131–53.

Beaulieu A. (2004) 'From brainbank to database: the informational turn in the study of the brain' in *Studies in History and Philosophy of Biological and Biomedical Sciences*, 35: 367–90.

Bennett J.M. and Kendrew J. (1952) 'The computation of Fourier syntheses with a digital electronic calculating machine' in *Acta Crystallographica*, 5: 109–16.

Benson K., Maienschein J. and Rainger R. (1988) *The American Development of Biology* (Philadelphia: University of Pennsylvania Press).

Benson K., Maienschein J. and Rainger R. (1991) *The Expansion of American Biology* (New Brunswick: Rutgers).

Bergin T. (2006a) 'The origins of word processing software for personal computers: 1976–1985' in *Annals of the History of Computing*, 28(4): 32–47, special issue on the history of word processing.

Bergin T. (2006b) 'The proliferation and consolidation of word processing software' in *Annals of the History of Computing*, 28(4): 48–63, special issue on the history of word processing.

Bergmann M. and Niemann C. (1938) 'On the structure of silk fibroin' in *Journal of Biological Chemistry*, 122: 577–96.

Beurton P., Falk R. and Rheinberger H.J. (eds, 2000) *The Concept of the Gene in Development and Evolution: Historical and Epistemological Perspectives* (Cambridge: Cambridge University Press).

Billeter M.A., Dahlberg J.E., Goodman H.M., Hindley J. and Weissmann C. (1969) 'Sequence of the first 175 nucleotides from the 5′ terminus of Qβ RNA synthesised in vitro' in *Nature*, 224: 1083–86.

Bodmer W. and McKie R. (1997) *The Book of Man: The Human Genome Project and the Quest to Discover our Genetic Heritage* (Oxford: Oxford University Press).

Bonneuil C. and Gaudillière J.P. (2007) 'Navigating the post-Fordist DNA: network, regulations and variability in genomics and society', paper delivered at the biennial meeting of the International Society for the History, Philosophy and Social Studies of Biology (ISHPSSB), University of Exeter.

Bonneuil C. and Thomas F. (2009) *Gènes, pouvoirs et profits: recherche publique et regimes de production des savoirs de Mendel aux OGM* [*Genes, Powers and Profits: Regimes of Knowledge Production from Mendel to the GMOs*] (Versailles and Lausanne: Quae).

Bostanci A. (2004) 'Sequencing human genomes' in J.P. Gaudillière and H.J. Rheinberger (eds) *From Molecular Genetics to Genomics: The Mapping Cultures of Twentieth Century Genetics* (London and New York: Routledge): 158–79.

Bostanci A. (2010) 'A metaphor made in public' in *Public Understanding of Science*, 32(4): 467–88.

Bostanci A. and Calvert J. (2008) 'Invisible genomes: the genomics revolution and patenting practice' in *Studies in History and Philosophy of Biological and Biomedical Sciences*, 39: 109–19.

Brandt C. (2005) 'Genetic code, text, and scripture: Metaphors and narration in German molecular biology' in *Science in Context*, 18(4): 629–48.

Brenner S. (1957) 'On the impossibility of all overlapping triplet codes in information transfer from nucleic acids to proteins' in *Proceedings of the National Academy of Sciences*, 43: 687–94.

Brenner S. (1963a) 'Letter to Max Perutz' in W. Wood (ed.) (1988) *The nematode Caenorhabditis elegans* (New York: Cold Spring Harbor): X–XI.

Brenner S. (1963b) 'Excerpts from proposal to the Medical Research Council' in W. Wood (ed.) (1988) *The nematode Caenorhabditis elegans* (New York: Cold Spring Harbor): XI–XIII.

Brenner S. (1973) 'The genetics of behaviour' in *British Medical Bulletin*, 29(3): 269–71.

Brenner S. (1974) 'The genetics of *Caenorhabditis elegans*' in *Genetics*, 77: 71–94.

Brenner S. (1990) 'The human genome: the nature of the enterprise' in *Ciba Foundation Symposia*, 149:6–17.

Brenner S. (2001) *My Life in Science* (London: BioMed).

Brenner S., Jacob F. and Meselson M. (1961) 'Unstable intermediate carrying information from genes to ribosomes for protein synthesis' in *Nature*, 190: 576–81.

Bretscher M.S. and Travers A.A. (2003) 'Obituary: John Smith' in *The Independent*, November 27th edition: 22 (broadsheet format) or 67 (compact format).

Bromberg J.L. (1991) *The Laser in America, 1950–1970* (Cambridge: MIT Press).

Brown A. (2004) *In the Beginning was the Worm: Finding the Secrets of Life in a Tiny Hermaphrodite* (New York: Columbia University Press).

Brown A. (2007) *J.D. Bernal: The Sage of Science* (Oxford: Oxford University Press).

Brown N., Rappert B. and Webster A. (2000) *Contested Futures: A Sociology of Prospective Techno-Science* (Aldershot: Ashgate).

Brown N., Kraft A. and Martin P. (2006) 'The promissory pasts of blood stem cells' in *Biosocieties,* 1(3): 329–48.

Brownlee G.G., Sanger F. and Barrell B. (1968) 'The sequence of 5s ribosomal ribonucleic acid' in *Journal of Molecular Biology*, 34(3): 379–412.

Buchwald J. and Warwick A. (eds, 2001) *Histories of the Electron: The Birth of Microphysics* (Cambridge: MIT Press).

Bud R. (1993) *The Uses of Life: A History of Biotechnology* (Cambridge: Cambridge University Press).

Bud R. (1998a) 'Molecular biology and the long-term history of biotechnology' in A. Thackray (ed.) *Private Science: Biotechnology and the Rise of the Molecular Sciences* (Philadelphia: University of Pennsylvania Press): 3–19.

Bud R. (1998b) 'Penicillin and the new Elizabethans' in *British Journal of the History of Science,* 31: 305–33.

Bud R. (2009) *Penicillin: Triumph and Tragedy* (Oxford: Oxford University Press).

Cairns J., Stent G. and Watson J. (1966) *Phage and the Origins of Molecular Biology* (New York: Cold Spring Harbor).

Caltech (1975–88) *Biology Division Annual Reports* (Pasadena: California Institute of Technology).

Campbell-Kelly M. (2003) *From Airline Reservations to Sonic the Hedgehog: A History of the Software Industry* (Cambridge: MIT Press).

Campbell-Kelly M. and Aspray W. (1996) *Computer: A History of the Information Machine* (New York: Basic Books).

Cantor C. and Smith C.L. (1999) *Genomics: The Science and Technology behind the Human Genome Project* (New York: Wiley Interscience).

Castells M. (2000 [1996]) *The Information Age: Economy Society and Culture* (Malden and Oxford: Blackwell Publishing), three vols.

Ceruzzi P. (1998) *A History of Modern Computing* (Cambridge: MIT Press).

Ceruzzi P. (2005) 'Moore's Law and technological determinism: reflections on the history of technology' in *Technology and Culture*, 46(3): 584–93.

Chandler Jr. A. (1977) *The Visible Hand: The Managerial Revolution in American Business* (Cambridge: Harvard University Press).

Chandler Jr. A. (2005) *Shaping the Industrial Century: The Remarkable Story of the Evolution of the Modern Chemical and Pharmaceutical Industries* (Cambridge: Harvard University Press).

Chang H. and Jackson C. (eds, 2007) *An Element of Controversy: The Life of Chlorine in Science, Technology, Medicine and War* (London: British Society for the History of Science).

Chargaff E. (1978) *Heraclitean Fire: Sketches from a Life before Nature.* (New York: Rockefeller University Press).

Chataway J., Tait J. and Wield D. (2004) 'Understanding company R&D strategies in agro-biotechnology: trajectories and blind spots' in *Research Policy*, 33(6–7): 1041–57. Reprinted in M. McKelvey and L. Orsenigo (eds, 2006) *The Economics of Biotechnology* (Cheltenham and Northampton: Edward Elgar), vol. 1: 328–44.

Chiang H. (2009) 'The laboratory technology of discrete molecular separation: the historical development of gel electrophoresis and the material epistemology of biomolecular science, 1945–1970' in *Journal of the History of Biology*, 42(3): 495–527.

Chibnall C. (1942) 'Amino-acid analysis and the structure of proteins' in *Proceedings of the Royal Society of London*, B series, 131: 136–60.

Chow-White P. (2007) *The Informationalisation of Race: Communication Technologies and Genomics in the Information Age.* PhD dissertation, Annenberg School for Communication, University of Southern California.

Chow-White P. and García-Sancho M. (2012) 'Bidirectional shaping and spaces of convergence: interactions of biology and computing from the first DNA sequencers to global genome databases' in *Science, Technology and Human Values*, 37(1): 124–64.

Clarke H. (1944) 'Obituary: Max Bergmann, 1886–1944' in *Science*, 102: 168–70.

CODASYL Systems Committee (1969) *A Survey of Generalized Data Base Management Systems* (New York: CODASYL).

Codd E.F. (1970) 'A relational model of data for large shared data banks' in *Communications of the ACM*, 13: 377–87.

Codd E.F. (1990) *The Relational Model for Database Management* (London: Addison Wesley).

Collins F. (2010) *The Language of Life: DNA and the Revolution in Personalised Medicine* (New York: Harper).

Collins S. (2004) *The Race to Commercialise Biotechnology: Molecules, Markets and the State in the United States and Japan* (London and New York: Routledge).

Consden R., Gordon A. and Martin A. (1944) 'Qualitative analysis of proteins: a partition chromatographic method using paper' in *Biochemical Journal*, 38: 224–32.

Consden R., Gordon A., Martin A. and Synge R. (1947) 'Gramicidin S: the sequence of the amino-acid residues' in *Biochemical Journal*, 41(4): 596–602.

Cook-Deegan R. (1994) *The Gene Wars: Science, Politics and the Human Genome* (New York: Norton).

Cormen T.H., Leiserson C. E., Rivest R. L. and Stein C. (2001) *Introduction to Algorithms* (Cambridge: MIT Press, 2nd edition).

Coulson A., Sulston J., Brenner S. and Karn J. (1986) 'Toward a physical map of the genome of the nematode *Caenorhabditis elegans*' in *Proceedings of the National Academy of Sciences of the US*, 83: 7821–5.

Creager A. (1998) 'Producing molecular therapeutics from human blood: Edwin Cohn's wartime enterprise' in S. de Chadarevian and H. Kamminga (eds) *Molecularizing Biology and Medicine: New Practices and Alliances, 1910s–1970s* (London: Harwood)): 99–128.

Creager A. and Gaudillière J.P. (1996) 'Meanings in search of experiments and vice-versa: the invention of allosteric regulation in Paris and Berkeley, 1959–1968' in *Historical Studies in the Physical and Biological Sciences*, 27: 1–89.

Creager A. and Santesmases M.J. (2006) 'Radiobiology in the atomic age: Changing research practices and policies in comparative perspective' in *Journal of the History of Biology*, 39: 637–47.

Crick F. (1958) 'On protein synthesis' in *Symposium of the Society for Experimental Biology*, 12: 138–63.

Crick F. (1967) 'Project K: the complete solution of *E. coli*' in *Perspectives in Biology and Medicine*, 17: 67–70.

Crick F. (1973) 'Project K: the complete solution of "E. coli"' in *Perspectives in Biology and Medicine*, 17: 67–70.

Crick F. (1988) *What Mad Pursuit: A Personal View of Scientific Discovery* (New York: Basic Books).

Danzin A. (1979) *Science and the Second Renaissance in Europe* (Brussels: Commission of the European Communities).

Dart E.C. (1989) 'Exploitation of biotechnology in a large company' in *Philosophical Transactions of the Royal Society of London*, B series, 324: 599–611.

Date C.J. (1981 [1975]) *An Introduction to Database Systems* (London: Addison Wesley).

Date C.J. (1990 [1987]) *Relational Database Writings, 1985–1989* (London: Addison Wesley).

Davies K. (2002) *Cracking the Genome: Inside the Race to Unlock Human DNA* (Baltimore: Johns Hopkins University Press).

Dayhoff M. and Ledley R. (1962) 'Comprotein: a computer program to aid primary protein structure determination' in *Proceedings of the Fall Joint Computer Conference* (Santa Monica: American Federation of Information Processing Societies): 262–74.

Dayhoff M. (1964) 'Computer aids to protein sequence determination' in *Journal of Theoretical Biology,* 8: 97–112.

De Chadarevian S. (1996) 'Sequences, conformation, information: biochemists and molecular biologists in the 1950s' in *Journal of the History of Biology*, 29: 361–86.

De Chadarevian S. (1998a) 'Following molecules: hemoglobin between the clinic and the laboratory' in S. De Chadarevian and H. Kamminga (eds) (1998) *Molecularizing Biology and Medicine: New Practices and Alliances, 1910s–1970s* (London: Harwood): 171–201.

De Chadarevian (1998b) 'Of worms and programmes: *Caenorhabditis elegans* and the study of development' in *Studies in History and Philosophy of Biological and Biomedical Sciences*, 29(1): 81–105.

De Chadarevian S. (1999) 'Protein sequencing and the making of molecular genetics' in *Trends in Biochemical Sciences*, 24: 203–6.

De Chadarevian S. (2000) 'Mapping development or how molecular is molecular biology?' in *History and Philosophy of the Life Sciences*, 22(3): 381–96.

De Chadarevian S. (2002) *Designs for Life: Molecular Biology after World War II* (Cambridge: Cambridge University Press).

De Chadarevian S. (2004) 'Mapping the worm's genome: tools, networks, patronage' in H.J. Rheinberger and J.P. Gaudillière (eds) *From Molecular Genetics to Genomics: The Mapping Cultures of Twentieth Century Genetics* (London and New York: Routledge): 95–110.

De Chadarevian S. (2009) 'Viewing chromosomes' in S. Brauckmann, C. Brandt, D. Thieffry and G.B. Müller (eds) *Graphing Genes, Cells and Embryos* (Berlin: Max Planck Institute for the History of Science, preprint number 380): 57–62.

De Chadarevian S. and Gaudillière J.P. (1996) 'The tools of the discipline: biochemists and molecular biologists' in *Journal of the History of Biology*, 29: 327–30.

De Chadarevian S. and Kamminga H. (eds, 1998) *Molecularizing Biology and Medicine: New Practices and Alliances, 1910s–1970s* (London: Harwood).

Deutsch T.F. (1988) 'Medical applications of lasers' in *Physics Today*, 41(10): 56–63.

Dietrich B. (1992) 'Future perfect – thanks to Bill Gates' 12 million dollar endowment, scientist Leroy Hood continues his search for a new genetic destiny' in *Seattle Times*, Sunday February 9th edition.

Dietrich M. (1998) 'Paradox and persuasion: negotiating the place of molecular evolution within evolutionary biology' in *Journal of the History of Biology*, 31: 85–111.

Dingman C.W. and Peacock A.C. (1968) 'Analytical studies on nuclear ribonucleic acid using polyacrylamide gel electrophoresis' in *Biochemistry*, 7(2): 659–68.

Dogson M. (1990) *Celltech: The First Ten Years of a Biotechnology Company* (Brighton: University of Sussex).

Donelson J. and Wu R. (1972) 'Nucleotide sequence analysis of deoxyribonucleic acid' in *Journal of Biological Chemistry*, 247(14): 4661–8.

Dreyer W. (1976) 'Peptide or protein sequencing method and apparatus', US Patent number 4,065,412, filed 7 May 1976 and issued 27 December 1977.

Dumas J.P. and Ninio J. (1982) 'Efficient algorithms for folding and comparing nucleic acid sequences' in *Nucleic Acids Research*, 10(1): 197–206.

Edgerton D. (1987) 'Science and technology in British business history' in *Business History*, 29(4): 84–103.

Edgerton D. (1996) *Science, Technology and the British Industrial 'Decline', 1870–1970* (Cambridge: Cambridge University Press).

Edgerton D. (1999) 'From innovation to use: ten eclectic theses on the historiography of technology' in *History and Technology*, 16: 111–36. French version available at *Annales HSS*, vols. 4–5, 1998.

Edgerton D. (2006) *The Shock of the Old: Technology and Global History Since 1900* (Oxford: Oxford University Press).

Edman P. (1949) 'A method for the determination of amino acid sequence in peptides' in *Archives of Biochemistry*, 22(3): 475.

Edman P. (1950) 'Method for determination of the amino acid sequence in peptides' in *Acta Chemica Scandinavica*, 4: 283–93.

Edman P. and Begg G. (1967) 'A protein sequenator' in *European Journal of Biochemistry*, 1: 80–91.

Edsall J.T. (1979) 'The development of the physical chemistry of proteins, 1898–1940' in *Annals of the New York Academy of Sciences*, 325: 53–76. Special issue on *The Origins of Modern Biochemistry: A Retrospect on Proteins*.

Erlich H., Gelfand D. and Sninski J.J. (1991) 'Recent advances in the polymerase chain reaction' in *Science*, 252: 1643–51.

Etxeberria A. and Umerez J. (2006) 'Organismo y Organización en la Biología Teórica: ¿Vuelta al organicismo?' ['Organism and organization in Theoretical Biology: back to organicism?'] in *Ludus Vitalis* XIV(26) : 3–38. Abstract available in English.

European Molecular Biology Laboratory (1977–1987) *Annual Reports* (Heidelberg: European Molecular Biology Laboratory).

Ferry G. (1998) *Dorothy Hodgkin: A Life* (London: Granta).

Fiers W., Contreras R., Duerinck F., Haegeman G., Iserentant D., Merregaert J., Min Jou W., Molemans F., Raeymaekers A., Van den Berghe A., Volckaert G. and Ysebaert M. (1976) 'Complete nucleotide-sequence of bacteriophage MS2-RNA: primary and secondary structure of replicase gene' in *Nature*, 260: 500–7.

Fischer E. (1902) 'Über die Hydrolyse der Proteïnstoffe' in *Chemiker Zeitung*, 26: 939–40.

Fischer E. (1907) 'Synthetical chemistry in its relation to biology' in *Transactions of the Chemical Society*, 91: 1749–65.

Fletcher L. and Porter R. (1997) *A Quest for the Code of Life: Genome Analysis at the Wellcome Trust Genome Campus* (London: Wellcome Trust).

Florkin M. (1972) *A History of Biochemistry* in Florkin M. and Stotz E. (eds) *Comprehensive Biochemistry*, vols 30–32 (New York: Elsevier).

Fortun M. (1993) *Mapping and Making Genes and Histories: The Genomics Project in the United States, 1980–1990*. PhD dissertation, Department of History of Science, Harvard University.

Fortun M. (1998) 'The Human Genome Project and the acceleration of biotechnology' in A. Thackray (ed.) *Private Science: Biotechnology and the Rise of the Molecular Sciences* (Philadelphia: University of Pennsylvania Press): 182–201.

Fortun M. (1999) 'Projecting high speed genomics' in Fortun M. and Mendelsohn E. (eds) *The Practices of Human Genetics* (Dordrecht: Kluwer): 25–48.

Fortun M. (2008) *Promising Genomics* (Berkeley: University of California Press).

Fox Keller E. (1995) *Refiguring Life: Changing Metaphors in Twentieth Century Biology* (New York: Columbia University Press).

Fox Keller E. (2000) *The Century of the Gene* (Cambridge: Harvard University Press).

Francoeur E. and Segal J. (2004) 'From model kits to interactive computer graphics' in S. de Chadarevian and N. Hopwood (eds) *Models: The Third Dimension of Science* (Standford: Stanford University Press): 402–29.

Frangione B. and Milstein C. (1968) 'Variations in the S-S bridges of immunoglobins G: interchain disulphide bridges of γG3 myeloma proteins' in *Journal of Molecular Biology*, 33: 893–906.

Fruton J. (1972) *Molecules and Life: Historical Essays on the Interplay of Chemistry and Biology* (New York: Wiley Interscience).

Fruton J. (1977) 'Early theories of protein structure' in *Annals of the New York Academy of Sciences*, 325: 1–20. Special issue on *The Origins of Modern Biochemistry: A Retrospect on Proteins*.

Fruton J. (1979) 'Early theories of protein structure' in *Annals of the New York Academy of Sciences*, 325: 1–20. Special issue on *The Origins of Modern Biochemistry: A Retrospect on Proteins*.

Fruton J. (1985) 'Contrasts in scientific style. Emil Fischer and Franz Hofmeister: their research groups and their theory of protein structure' in *Proceedings of the American Philosophical Society*, 129(4): 313–70.

Fruton J. (1992) *A Skeptical Biochemist* (Cambridge: Harvard University Press).

Fruton J. (1999) *Proteins, Enzymes and Genes: The Interplay of Chemistry and Biology* (New Haven: Yale University Press).

Fujimura J. (1999) 'The practices of producing meaning in bioinformatics' in M. Fortun and E. Mendelsohn (eds) *The Practices of Human Genetics* (Dordrecht: Kluwert): 49–88.

Fujimura J. and Fortun M. (1996) 'Constructing knowledge across social worlds: the case of DNA sequence databases in molecular biology', in L. Nader (ed.) *Naked Science: Anthropological Inquiry into Boundaries, Power, and Knowledge* (New York and London: Routledge): 160–73.

Fung S., Woo S. and Smith L. (1985) 'Amino derivatised phosphite and phosphate linking agents, phosphoramidite precursors, and useful conjugates thereof'. US Patent number 4,757,141, filed 26 August 1985 and issued 12 July 1988.

Galison P. and Hevly B. (eds, 1992) *Big Science: The Growth of Large-Scale Research* (Stanford: Stanford University Press).

Gannett L. (2004) 'The biological reification of race' in *British Journal for the Philosophy of Science*, 55(2): 323–45.

Gannett L. and Griesemer J. (2004) 'Classical genetics and the geography of genes' in J.P. Gaudillière and H.J. Rheinberger (eds) *Classical Genetic Research and Its Legacy: The Mapping Cultures of Twentieth Century Genetics* (London and New York: Routledge): 57–87.

García-Sancho M. (2007a) 'The rise and fall of the idea of genetic information (1948–2006)' in *Genomics, Society and Policy*, 2(3): 16–36.

García-Sancho M. (2007b) 'Mapping and sequencing information: the social context for the genomics revolution' in *Endeavour*, 31(1): 18–23.

García-Sancho M. (2008) *Sequencing as a Way of Work: A History of Its Emergence and Mechanisation – From Proteins to DNA, 1945–2000*. PhD dissertation, Centre for the History of Science, Technology and Medicine, Imperial College, London.

García-Sancho M. (2009) 'The perception of an information society and the emergence of the first computerised biological databases' in A. Matsumoto and N. Nakano (eds) *Human Genome. Features, Variations and Genetic Disorders* (New York: Nova Science Publishers): 257–76.

García-Sancho M. (2010) 'A new insight into Sanger's development of sequencing: from proteins to DNA, 1943–1977' in *Journal of the History of Biology*, 43(2): 265–323.

García-Sancho M. (2011a) 'Academic and molecular matrices: a study of the transformations of connective tissue research at the University of Manchester (1947–1996)' in *Studies in History and Philosophy of Biological and Biomedical Sciences*, 42(2): 233–45.

García-Sancho M. (2011b) 'From metaphor to practices: the introduction of information engineers into the first DNA sequence database' in *History and Philosophy of the Life Sciences*, 33(1): 71–104.

García-Sancho M. (2012) 'From the genetic to the computer program: the historicity of "data" and "computation" in the investigations on the nematode worm *C. elegans* (1963–1998)' in *Studies in History and Philosophy of Biological and Biomedical Sciences,* 43: 16–28. Special issue on *Data-Driven Science.*

Garesse R. and Arribas E. (1987) 'Secuenciación de DNA y síntesis de oligonucleótidos' ['DNA sequencing and oligonucleotide synthesis'] in M. Vicente and J. Renart (eds) *Ingeniería Genética* [*Genetic Engineering*] (Madrid: Consejo Superior de Investigaciones Científicas): 71–90.

Gaudillière J.P. (1992) 'J. Monod, S. Spiegelman et l'adaptation enzymatique: programmes de recherche, cultures locales et traditions disciplinaires.' ['J. Monod, S. Spiegelman and enzymatic adaptation: research programmes, local cultures and disciplinary traditions'] in *History andPhilosophy of the Life Sciences,* 14: 29–98.

Gaudillière J.P. (1993) 'Molecular biology in the French tradition? Redefining local traditions and disciplinary patterns' in *Journal of the History of Biology,* 26(3): 473–98.

Gaudillière J.P. (1996) 'Molecular biologists, biochemists and messenger RNA: the birth of a scientific network' in *Journal of the History of Biology,* 29(3): 417–45.

Gaudillière J.P. (2002) *Inventer la biomédicine: la France, l'Amérique et la production des savoirs du vivant (1945–1965)*[*Inventing Biomedicine: France, America and the Production of Knowledge about Living Entities*] (Paris: La Découverte).

Gaudillière J.P. (2004a) 'Genesis and development of a biomedical object: styles of thought, styles of work and the history of the sex steroids' in *Studies in History and Philosophy of Biological and Biomedical Sciences,* 35(3): 525–43.

Gaudillière J.P. (2004b) 'Mapping as technology: genes, mutant mice, and biomedical research' in Gaudillière J.P. and H.J. Rheinberger (eds) *Classical Genetic Research and its Legacy: The Mapping Cultures of Twentieth Century Genetics* (London and New York: Routledge): 173–203.

Gaudillière J.P. (2009) 'New wine in old bottles? The biotechnology problem in the history of molecular biology' in *Studies in History and Philosophy of Biological and Biomedical Sciences,* 40: 20–8.

Gaudillière J.P. and Hess V. (eds, 2009) *Ways of Regulating: Therapeutic Agents between Plants, Shops and Consulting Rooms* (Berlin: Max Planck Institute for the History of Science Preprint Series, number 363).

Gay H. (2007) *The history of Imperial College London 1907–2007 : Higher Education and Research in Science, Technology and Medicine* (London: Imperial College Press).

Gee H. (2004) *Jacob's Ladder: The History of the Human Genome* (New York: Norton).

Gee S. (2007) *The Case of Research Tools for Drug Discovery: Evolving User-Producer Inter-Dependencies and the Exchange of Knowledge.* PhD dissertation, Manchester Business School, University of Manchester.

Geison G. (1978) *Michael Foster and the Cambridge School of Physiology* (Princeton: Princeton University Press).

Gilbert W. (1980) 'DNA sequencing and gene structure'. Nobel Lecture, available at www.nobel.se/chemistry/laureates/1980/gilbert-lecture.html, last accessed January 2012.

Gilbert W. (1992) 'A vision of the grail' in D.J. Kelves and L. Hood (eds) *The Code of Codes: Scientific and Social Issues in the Human Genome Project* (Cambridge: Harvard University Press).

Gingeras T., Milazzo J., Sciaky D. and Roberts R. (1979) 'Computer programs for the assembly of DNA sequences' in *Nucleic Acids Research*, 7(2): 529–43.

Gingeras T.R. and Roberts R. (1980) 'DNA sequence data analysis: steps toward computer analysis of nucleotide sequences' in *Science*, 209: 1322–8.

Gingeras T.R., Rice P. and Roberts R.J. (1982) 'A semi-automated method for the reading of nucleic acid sequencing gels' in *Nucleic Acids Research*, 10(1): 103–14.

Gingras Y. (2010) 'Revisiting the "quiet debut" of the double helix: A bibliometric and methodological note on the "impact" of scientific publications' in *Journal of the History of Biology*, 43: 159–181.

Godet M. and Ruyssen O. (1979) *The Old World and the New Technologies* (Brussels: Commission of the European Communities).

Goodstein J. (1991) *Millikan's School: A History of the California Institute of Technology* (New York: Norton).

Gordon A.H. (1977) 'Electrophoresis and chromatography of amino acids and proteins' in *Annals of the New York Academy of Sciences*, 325: 95–106. Special issue on *The Origins of Modern Biochemistry: A Retrospect on Proteins*.

Gordon A.H., Martin A. and Synge R. (1943) 'The amino acid composition of gramicidin' in *Biochemical Journal*, 37(1): 86–92.

Gosselin P. and Jacobs P. (2000) 'DNA device's heredity scrutinised by US' in *Los Angeles Times*, Sunday May 14th edition: A18–A19.

Gossling T.H. and Mallett W.F. (1968) 'Use of computers in a molecular biology laboratory' in *Proceedings of the Fall Joint Computer Conference* (Santa Monica: American Federation of Information Processing Societies): 1089–98.

Gottweis H. (1998a) *Governing Molecules: The Discursive Politics of Genetic Engineering in Europe and the United States* (Cambridge: MIT Press).

Gottweis H. (1998b) 'The political economy of British biotechnology' in A. Thackray (ed.) *Private Science: Biotechnology and the Rise of the Molecular Sciences* (Philadelphia: University of Pennsylvania Press): 105–30.

Griffiths P. (2001) 'Genetic information: a metaphor in search of a theory' in *Philosophy of Science*, 68: 394–412.

Griffiths P.E. and Stotz K. (2006) 'Genes in the postgenomic era' in *Theoretical Medicine and Bioethics*, 27(6): 499–521.

Griswold R.E. (1978) 'A history of the SNOBOL programming languages' in R.L. Wexelblat (ed.) *History of Programming Languages I* (New York: ACM): 601–45.

Guenther A.H., Kressel, H.R. and Krupke, W.F. (1991) 'The laser now and in the future' in J. L. Bromberg (ed.) *The Laser in America, 1950–1970* (Cambridge: MIT Press): 228–48.

Guston D. and Keniston, K. (1994) 'The social contract for science' in Guston D. and Keniston, K. (eds) *The Fragile Contract: University Science and the Federal Government* (Cambridge: MIT Press): 1–41.

Hagen J. (1999) 'Naturalists, molecular biologists and the challenges of molecular evolution' in *Journal of the History of Biology*, 32: 321–34.

Hagen J. (2001) 'The introduction of computers into systematic research in the United States during the 1960s' in *Studies in History and Philosophy of Biological and Biomedical Sciences*, 32(2): 291–314.

Hagen J. (2010) 'Waiting for sequences: Morris Goodman, immunodiffusion experiments, and the origins of molecular anthropology' in *Journal of the History of Biology*, 43(4): 697–725.

Haigh T. (2001) 'Inventing information systems: the systems men and the computer, 1950–1968' in *Business History Review,* 75(1): 15–61.

Haigh T. (2006a) 'A veritable bucket of facts: origins of the data base management system' in *ACM SIGMOD Record,* 35(2): 33–49.

Haigh T. (2006b) 'Remembering the office of the future: the origins of word processing and office automation' in *Annals of the History of Computing,* 28(4): 6–31, Special issue on the history of word processing.

Hamilton D. (1991) 'Leroy Hood to move north' in *Science,* 254: 189.

Hamm G. and Stüber, K. (1982) 'The European Molecular Biology Laboratory (EMBL) nucleotide sequence data library' in *Nucleotide Sequence Data Library News,* 1: 2–8.

Hamm G. and Cameron, G. (1986) 'The EMBL Data Library' in *Nucleic Acids Research,* 14(1): 5–9.

Haraway D. (1976) *Crystals, Fabrics and Fields: Metaphors of Organicism in Twentieth-Century Developmental Biology* (New Haven: Yale University Press).

Haraway D. (1997) *Modest_Witness@Second_Millennium. FemaleMan_Meets_ OncoMouse: Feminism and Technoscience* (New York and London: Routledge).

Harper P. (2008) *A Short History of Medical Genetics* (Oxford: Oxford University Press).

Hartley B. (1970a) 'The primary structure of proteins' in T.W. Goodwin (ed.) *British Biochemistry: Past and Present,* Biochemical Society Symposia, n° 30 (London: Academic Press): 29–41.

Hartley B. (1970b) 'Strategy and tactics in protein chemistry' in *Biochemical Journal,* 119: 805–22.

Hartley B. (2004) 'The first floor, Department of Biochemistry, University of Cambridge' in *IUBMB Life,* 56(7): 437–9.

Hartley B., Brown J., Kauffman D. and Smillie L. (1965) 'Evolutionary similarities between pancreatic proteolitic enzymes' in *Nature,* 207: 1157–9.

Harvey M. and McMeekin A. (2007) *Public or Private Economies of Knowledge? Turbulence in the Biological Sciences* (Cheltenham and Northampton: Edward Elgar).

Harwood J. (1993) *Styles of Scientific Thought: The German Genetics Community* (Chicago: University of Chicago Press).

Hayles N.K (1999) *How We Became Posthuman: Virtual Bodies in Cybernetics, Literature, and Informatics* (Chicago: University of Chicago Press).

Häyry M., Chadwick R. Árnason V. and Árnason G. (2007) *The Ethics and Governance of Human Genetic Databases: European Perspectives* (Cambridge: Cambridge University Press).

Heeney C. and Smart A. (2012) 'Enacting governance – the case of access' in J. Kaye (ed.) *Governing Biobanks: Understanding the Interplay between Law and Practice* (Oxford: Hart Publishers): 232–58.

Hendry J. (ed., 1984) *Cambridge Physics in the Thirties* (Abingdon: Taylor & Francis).

Herran N. and Roqué X. (2009) 'Tracers of modern technoscience' in *Dynamis,* 29: 123–30.

Hewick R., Hunkapiller M., Hood L. and Dreyer W. (1981) 'A gas-liquid solid phase peptide and protein sequenator' in *Journal of Biological Chemistry,* 256(15): 7790–7.

Hilgartner S. (2004) 'Making maps and making social order: governing American genome centers, 1988–93' in H.J. Rheinberger and J.P. Gaudillière (eds) *The Mapping Cultures of Twentieth Century Genetics* (London and New York: Routledge): 113–28.

Hilgartner S. (2008) 'Laboratory by any other name? Factory discourse in genomic research', paper presented at the workshop *Writing Genomics: Historiographical Challenges for New Historical Developments*, Max Planck Institute for the History of Science, Berlin.

Hirs C., Moore S. and Stein W. (1960) 'The sequence of the amino acid residues in performic acid-oxidized ribonuclease' in *Journal of Biological Chemistry*, 235(3): 633–47.

Hofmeister F. (1902) 'Über Bau und Gruppierung der Eiweisskörper' in *Ergebnisse der Physiologie*, 1: 759–802.

Hofschneider P. and Murray K. (2001) 'Combining science and business: from recombinant DNA to vaccines against hepatitis B virus' in P. Buckel (ed.) *Recombinant Protein Drugs* (Basel, Boston and Berlin: Birkhäuser): 43–64.

Holley R.W., Apgar J., Everett G.A., Madison J.T., Marquisee M., Merrill S.H., Penswick J.R. and Zamir A. (1965) 'Structure of a ribonucleic acid' in *Science* 147: 1462–5.

Holmes F.L. (2001) *Meselson, Stahl and the Replication of DNA: A History of 'The Most Beautiful Experiment in Biology'* (New Haven: Yale University Press).

Holmes F.L. (2006) *Reconceiving the Gene: Seymour Benzer's Adventures in Phage Genetics* (New Haven: Yale University Press).

Holmes F.L. and Heilbron J.L. (1992) *Between Biology and Medicine: The Formation of Intermediary Metabolism* (Berkeley: University of California).

Hood L. (1990) 'Human Genome Project: is big science bad for biology? No, and anyway the HGP isn't big science' in *The Scientist*, 4(22): 13–15.

Hood L. (1992) 'Biology and medicine in the twenty first century' in D.J. Kevles and L. Hood (eds) *The Code of Codes: Scientific and Social Issues in the Human Genome Project* (Cambridge: Harvard University Press): 136–63.

Hood L. (2003) 'A personal view of molecular technology and how it has changed biology' in *Journal of Proteome Research*, 1(5): 399–409.

Hood L. (2008) 'A personal journey of discovery: developing technology and changing biology' in *Annual Review of Analytical Chemistry*, 1: 1.1–1.43.

Hood L. and Hunkapiller M. (1979) 'Apparatus for the performance of chemical processes'. US Patent number 4,252,769, filed 26 December 1979 and issued 24 February 1981.

Hood L, Huang, H. and Dreyer, W. (1977) 'The area code hypothesis: the immune system provides clues to understanding the genetic and molecular basis of cell recognition during development' in *Journal of Supramolecular Structure*, 7(3–4): 531–59.

Hood L. and Smith L. (1987) 'Genome sequencing: how to proceed' in *Issues in Science and Technology*, 3(3): 36–46.

Hotta Y. and Benzer S. (1972) 'Mapping of behaviour in *Drosophila* mosaics' in *Nature*, 240: 527–35.

Howlett P. and Morgan M. (eds, 2010) *How Well Do Facts Travel?* (Cambridge: Cambridge University Press).

Hughes T.P. (1989) *American Genesis: A Century of Invention and Technological Enthusiasm, 1870–1970* (New York: Viking).

Hughes T.P. (1990) *American Genesis: A Century of Invention and Technological Enthusiasm* (London: Penguin).

Hunkapiller T., Kaiser R.J., Koop B.F. and Hood L. (1991) 'Large-scale and automated DNA sequence determination' in *Science*, 254: 59–67.

Hutchison III C.A. (2007) 'DNA sequencing: bench to bedside and beyond' in *Nucleic Acids Research*, 35(18): 6227–37.

Ingram V. (1958) 'Abnormal human haemoglobins: I. The comparison of normal human and sickle-cell haemoglobins by fingerprinting' in *Biochimica et Biophysica Acta*, 28: 539–45.

Isono K. (1982) 'Computer programs to analyse DNA and amino acid sequence data' in *Nucleic Acids Research*, 10(1): 85–9.

Jacob F. (1974) *The Logic of Living Systems: A History of Heredity* (London: Allen Lane).

Jeppesen P., Barrell B., Sanger F. and Coulson A. (1972) 'Nucleotide sequence of two fragments from the coat-protein cistron of bacteriophage R17 ribonucleic acid' in *Biochemical Journal*, 128(5): 993–1006.

Jordan K. and Lynch M. (1993) 'The dissemination, standardization and routinization of a molecular biological technique' in *Social Studies of Science*, 28(5–6): 773–800.

Judson H.F. (1979) *The Eight Day of Creation: Makers of the Revolution in Biology* (London: Cape).

Judson H.F. (1992) 'A history of the science and technology behind gene mapping and sequencing' in D.J. Kevles and L. Hood (eds) *The Code of Codes: Scientific and Social Issues in the Human Genome Project* (Cambridge: Harvard University Press): 37–80.

Kamminga H. and Cunningham A. (eds, 1995) *The Science and Culture of Nutrition, 1840–1940* (Amsterdam: Rodopi).

Kamminga H. and Weatherall M.W. (1996) 'The making of a biochemist. I: Frederick Gowland Hopkins' construction of dynamic biochemistry' in *Medical History*, 40(3): 269–92.

Kanehisa M., Klein P., Greif P. and DeLisi C. (1984) 'Computer analysis and structure prediction of nucleic acids and proteins' in *Nucleic Acids Research*, 12(1, Part 1): 417–28.

Kass L. and Bonneuil C. (2004) 'Mapping and seeing: Barbara McClintock and the linking of genetics and cytology in maize genetics, 1928–35' in J.P. Gaudillière and H.J. Rheinberger (eds) *Classical Genetic Research and its Legacy: The Mapping Cultures of Twentieth Century Genetics* (London: Routledge): 91–118.

Kaufmann A. (2004) 'Mapping the human genome at Généthon laboratory' in J.P. Gaudillière and H.J. Rheinberger (eds) *From Molecular Genetics to Genomics: The Mapping Cultures of Twentieth Century Genetics* (London and New York: Routledge): 129–57.

Kay L. (1988) 'Laboratory technology and biological knowledge: the Tiselius electrophoresis apparatus, 1930–1945' in *History and Philosophy of the Life Sciences*, 10(1): 51–72.

Kay, L. (1993) *The Molecular Vision of Life: Caltech, the Rockefeller Foundation and the Rise of the New Biology* (New York: Oxford University Press).

Kay L. (2000) *Who Wrote the Book of Life: A History of the Genetic Code* (Stanford: Stanford University Press).

Kaye J. (ed., 2012) *Governing Biobanks: Understanding the Interplay between Law and Practice* (Oxford: Hart Publishers).

Keating P., Limoges C. and Cambrosio A. (1999) 'The automated laboratory: the generation and replication of work in molecular genetics' in M. Fortun and E. Mendelsohn (eds) *The Practices of Human Genetics* (Dordrecht: Kluwer): 125–42.

Keating P. and Cambrosio A. (2006) *Biomedical Platforms: Realigning the Normal and the Pathological in Late-Twentieth-Century Medicine* (Cambridge: MIT Press).

Kennard O. (1997) 'From private data to public knowledge' in *The Impact of Electronic Publishing on the Academic Community*, International Workshop Organised by the Academia Europaea and the Wenner-Gren Foundation. Available on-line at www.portlandpress.com/pp/books/online/tiepac/session6/ch2.htm, last accessed January 2012.

Kenney M. (1986) *Biotechnology: The University-Industrial Complex* (New Haven: Yale University Press).

Kernighan B. and Plauger P. (1976) *Software Tools* (London: Addison-Wesley).

Kevles D.J. (1985) *In the Name of Eugenics: Genetics and the Uses of Human Heredity* (Cambridge: Harvard University Press).

Kevles D.J. and Hood L. (eds, 1992) *The Code of Codes: Scientific and Social Issues in the Human Genome Project* (Cambridge: Harvard University Press).

Kline R. (2006) 'Cybernetics, management science and technology policy' in *Technology and Culture*, 47: 513–35.

Kohara Y., Akiyama K. and Isono K. (1987) 'The physical map of the whole *E. coli* chromosome: application of a new strategy for rapid analysis and sorting of a large genomic library' in *Cell*, 50: 495–508.

Kohler R. (1978) 'Walter Fletcher, F. G. Hopkins, and the Dunn Institute of Biochemistry: a case study in the patronage of science' in *Isis*, 69: 331–55.

Kohler R. (1982) *From Medical Chemistry to Biochemistry* (Cambridge: Cambridge University Press).

Kohler R. (1994) *Lords of the Fly: Drosophila Genetics and the Experimental Life* (Princeton: Princeton University Press).

Kornberg A. (1987) 'The two cultures: chemistry and biology' in *Biochemistry*, 26: 6888–91.

Krige J. (2002) 'The birth of EMBO and the difficult road to EMBL' in *Studies in History and Philosophy of Biological and Biomedical Sciences*, 33: 547–64.

Kumar V., Sercarz E., Bressler J. and Laterra J. (1991) 'Hood lab investigation' in *Science*, 254: 1090–1.

Kuska B. (1998) 'Beer, Bethesda and biology: how genomics came into being' [interview with T.H. Roderick] in *Journal of the National Cancer Institute*, 90: 93.

Lagnado J. (2005) 'Questions and answers with Fred Sanger' in *The Biochemist*, 27(6): 37–9.

Lancelot G. (2007) *The Many Faces of Reform: The Reorganisation of Academic Life Sciences in Britain and France, 1965–1995*. PhD dissertation, Centre for the History of Science, Technology and Medicine, University of Manchester.

Latour B. (1987) *Science in Action: How to Follow Scientists and Engineers through Society* (Cambridge: Harvard University Press).

Lawn R., Fritsch E., Parker R., Blake G. and Maniatis, T. (1978) 'The isolation and characterisation of linked δ- and β-globin genes from a cloned library of human DNA' in *Cell*, 15: 1157–74.

Lécuyer C. (2006) *Making Silicon Valley: Innovation and the Growth of High Tech, 1930–1970* (Cambridge: MIT Press).

Lederberg J. and McCray, A. (2001) 'Ome sweet omics: a genealogical treasury of words' in *The Scientist*, 15: 8.

Ledley R. (1965) *Use of Computers in Biology and Medicine* (New York: McGraw Hill).

Lenoir T. (1999) 'Shaping biomedicine as an information science' in M.E. Bowden, T.B. Hahn and R.V. Williams (eds) *Proceedings of the 1998 Conference on the History and Heritage of Science Information Systems* (Medford: ASIS): 27–45.

Lenoir T. (2002a) 'Makeover: writing the body into the posthuman technoscape. Part one: embracing the posthuman', special issue of *Configurations*, 10(2): 203–20.

Lenoir T. (2002b) 'Makeover: writing the body into the posthuman technoscape. Part two: corporeal axiomatics', special issue of *Configurations*, 10(3): 373–85.

Lenoir T. and Hays M. (2000) 'The Manhattan Project for biomedicine' in P.R. Sloan (ed.) *Controlling our Destinies: Historical, Philosophical, Ethical and Theological Perspectives on the Human Genome Project* (Indiana: University of Notre Dame): 19–46.

Leonelli S. (2008) 'Bio-ontologies as tools for integration in biology' in *Biological Theory*, 3: 7–11.

Leonelli S. (2010a) 'Packaging small facts for reuse: databases in model organism biology' in P. Howlett and M. Morgan (eds) *How Well Do Facts Travel?* (Cambridge: Cambridge University Press): 325–48.

Leonelli S. (2010b) 'Documenting the emergence of bio-ontologies: or, why researching bioinformatics requires HPSSB' in *History and Philosophy of the Life Sciences*, 32(1): 105–26.

Leonelli S. (2012) 'Introduction: making sense of data-driven research in the biological and biomedical sciences', *Studies in History and Philosophy of Biological and Biomedical Sciences*, 43(1): 1–3 special issue on *Data-Driven Science*.

Leslie S.W. (1993) *The Cold War and American Science: The Military-Industrial-Academic Complex at MIT and Stanford* (New York: Columbia University Press).

Levitt M. (1969) 'Detailed molecular model for transfer ribonucleic acid' in *Nature*, 224: 759–63.

Levitt M. (2001) 'The birth of computational structural biology' in *Nature Structural Biology*, 8: 392–3.

Levitt M. and Warshel A. (1975) 'Computer simulation of protein folding' in *Nature*, 253: 694–8.

Lewin R. (1984) 'Why is development so illogical?' in *Science*, 224: 1327–9.

Liebenau J. (1987) 'The British success with penicillin' in *Social Studies of Science*, 17: 69–86.

Liebenau J. (1989) 'The MRC and the pharmaceutical industry: the model of insulin' in L. Bryder and J. Austoker (eds) *Historical Perspectives on the Role of the MRC: Essays in the History of the Medical Research Council of the United Kingdom and Its Predecesor, the Medical Research Committee, 1913–1953* (Oxford: Oxford University Press): 163–80.

Lindee S. (2005) *Moments of Truth in Genetic Medicine* (Baltimore: Johns Hopkins University Press).

López-Beltrán C. (2004) *El sesgo hereditario: ámbitos históricos del concepto de herencia biológica* [*The Hereditary Bias: Historical Settings of the Concept of Biological Inheritance*] (México: Universidad Nacional Autónoma de México).

Lynch M. and Woolgar S. (eds, 1988) *Representation in Scientific Practice* (Cambridge: MIT Press).

Lynch M., Cole S.A., McNally R. and Jordan K. (2009) *Truth Machine: The Contentious History of DNA Fingerprinting* (Chicago: University of Chicago Press).

MacKenzie D. and Wajcman J. (eds, 1999) *The Social Shaping of Technology* (Buckingham: Open University Press).

Maddox B. (2003) *Rosalind Franklin: The Dark Lady of DNA* (London: Harper Collins).

Martin A. (1952) 'The development of partition chromatography', Nobel Lecture, available at www.nobel.se/chemistry/laureates/1952/martin-lecture.html, last accessed January 2012.

Martin A. and Synge R. (1941) 'A new form of chromatogram employing two liquid phases' in *Biochemical Journal*, 35: 1358–68.

Matthaei H. and Nirenberg M. (1961) 'Characterisation and stabilisation of DNAase-sensitive protein synthesis in *E. coli* extracts' in *Proceedings of the National Academy of Sciences*, 47: 1580–8.

Maxam A. and Gilbert W. (1977) 'A new method for sequencing DNA' in *Proceedings of the National Academy of Sciences*, 74: 560–4.

Maxam A.M. and Gilbert W. (1980) 'Sequencing end-labeled DNA with base-specific chemical cleavages' in *Methods Enzymology*, 65(1):499–560.

Maynard Smith J. (2000) 'The concept of information in biology' in *Philosophy of Science*, 67: 177–94.

McAfee K. (2003) 'Neoliberalism on the molecular scale: economic and genetic reductionism in biotechnology battles' in *Geoforum,* 34: 203–19.

McCallum and Smith (1977) 'Computer processing of DNA sequence data' in *Journal of Molecular Biology*, 116: 29–30.

McKelvey and Orsenigo (eds, 2006) *The Economics of Biotechnology* (Cheltenham and Northampton: Edward Elgar), 2 vols.

McKusick V. (1966) *Mendelian Inheritance in Man* (Baltimore: Johns Hopkins University Press).

McKusick V. and Ruddle F. (1987) 'Editorial: a new discipline, a new name, a new journal', in *Genomics*, 1: 1–2.

McLachlan A.D. (1971) 'Tests for comparing related amino acid sequences. Cytochrome c and cytochrome c551' in *Journal of Molecular Biology,* 61: 409–24.

McLachlan A.D. (1976) 'Evidence for gene duplication in collagen' in *Journal of Molecular Biology*, 107: 159–74.

McLachlan A.D., Bloomer A.C. and Butler P.J.G. (1980) 'Structural repeats and evolution of Tobacco Mosaic Virus coat protein and RNA' in *Journal of Molecular Biology,* 136: 203–24.

Mendelsohn. E. (1992) 'The social locus of scientific instruments' in R. Bud and S.E. Cozzens (eds) *Invisible Connections: Instruments, Institutions and Science* (Bellingham: SPIE Optical Engineering Press): 5–22.

Mindell D. (2002) *Between Human and Machine: Feedback, Control and Computing before Cybernetics* (Baltimore: Johns Hopkins University Press).

Monod J. (1972) *Chance and Necessity: An Essay on the Natural Philosophy of Modern Biology* (London: Collins).

Moody G. (2004) *Digital Code of Life: How Bioinformatics is Revolutionising Science, Medicine and Business* (London: Wiley).

Moore S. and Stein W. (1972) 'The chemical structure of pancreatic ribonuclease and deoxiribonuclease', Nobel Lecture, available at www.nobel.se/chemistry/laureates/1972/moore-lecture.html, last accessed January 2012.

Morange M. (2000) *A History of Molecular Biology* (Cambridge: Harvard University Press).

Morange M. (2008) 'The death of molecular biology?' in *History and Philosophy of the Life Sciences*, 30(1): 31–42.

Morgan T.H., Bridges C.B. and Sturtevant A.H. (1925) 'The genetics of *Drosophila*' in *Bibliographica Genetica*, II: 1–262.

Morris P. (1998) 'Chromatograph' in R. Bud and D.J. Warner (eds) *Instruments of Science: An Historical Encyclopaedia* (New York and London: Garland Publishing): 107–110.

Moss L. (2004) *What Genes Can't Do* (Cambridge: MIT Press).

Mowery D. (1984) 'Firm structure, Government policy, and the organisation of industrial research: Great Britain and the United States, 1900–1950' in *Business History Review*, 58: 504–31. Reprinted in D. Edgerton (ed., 2006) *Industrial Research and Innovation in Business* (Cheltenham and Northampton: Edward Elgar).

Müller-Wille (2003) 'Joining Lapland and the Topinambes in flourishing Holland: centre and periphery in Linnaean botany' in *Science in Context*, 16(4): 461–88.

Müller-Wille S. and Rheinberger H.J. (eds, 2007) *Heredity Produced: At the Crossroads of Biology, Politics, and Culture, 1500–1870* (Cambridge: MIT Press)

Mullis K. (1990) 'The unusual origin of the polymerase chain reaction' in *Scientific American*, 262: 56–65.

Mullis K. (1994) 'PCR and scientific invention: the trial of DuPont vs. Cetus' in K. Mullis, F. Ferre and R. Gibbs (eds) *PCR: The Polymerase Chain Reaction* (Boston: Birkhäuser): 427–41.

Mullis K. (1998) *Dancing Naked in the Mind Field* (New York: Vintage).

Mullis K., Faloona F., Scharf S., Saiki R., Horn G. and Erlich H. (1986) 'Specific enzymatic amplification of DNA in vitro: the Polymerase Chain Reaction' in *Cold Spring Harbor Symposia on Quantitative Biology*, 51: 263–73.

Mullis K. and Faloona F. (1987) 'Specific synthesis of DNA *in vitro* via a polymerase-catalyzed chain reaction' in *Methods Enzymology*, 155: 335–50.

Murray K. (1970) 'Nucleotide maps of digests of deoxyribonucleic acids' in *Biochemical Journal*, 118: 831–41.

Nakamura L. and Chow-White P. (2011) *Race after the Internet* (New York and London: Routledge).

Needleman S. and Wunsch D. (1970) 'A general method applicable to the search for similarities in the amino acid sequence of two proteins' in *Journal of Molecular Biology*, 48: 443–53.

Neuberger A. (1939) 'Chemical criticism of the cyclol and frequency hypothesis of protein structure' in *Proceedings of the Royal Society of London*, B series, 127: 25–6.

Nirenberg M. and Matthaei H. (1961) 'The dependence of cell-free protein synthesis in E. coli upon naturally occurring or synthetic polyribonucleotides' in *Proceedings of the National Academy of Sciences*, 47: 1588–1602.

Noble D. (1979) *America by Design: Science, Technology and the Rise of Corporate Capitalism* (Oxford: Oxford University Press).

Noble D. (1984) *Forces of Production: A Social History of Industrial Automation* (New York: Knopf).

November J. (2004) 'LINC: biology's revolutionary little computer' in *Endeavour*, 28(3): 125–31.

November J. (2006) *Digitizing Life: The Introduction of Computers to Biology and Medicine*. PhD dissertation, Department of History of Science, Princeton University.

November J. (in press) *Digitizing Life: The Rise of Biomedical Computing in the United States* (Baltimore: Johns Hopkins University Press).

Office of Technology Assessment (1984) *Commercial Biotechnology: An International Analysis* (Washington: US Congress).

Office of Technology Assessment (1988) *Mapping Our Genes – Genome Projects. How Big? How Fast?* (Washington: US Congress).

Olazaran M. (1996) 'A sociological study of the official history of the perceptrons controversy' in *Social Studies of Science*, 26(3): 611–59.

Olby R. (1992 [1974]) *The Path to the Double Helix: The Discovery of DNA* (New York: Courier Dover Publications).

Olby R. (2009) *Francis Crick: Hunter of Life's Secrets* (New York: Cold Spring Harbor).

Olson M., Dutchik J.E., Graham M.Y., Brodeur G.M., Helms C., Frank M., MacCollin M., Scheinman R. and Frank T. (1986) 'Random-clone strategy for genomic restriction mapping in yeast' in *Proceedings of the National Academy of Sciences*, 83: 7826–30.

Onaga L. (2005) 'Ray Wu and DNA sequencing', paper delivered at the biennial meeting of the International Society for the History, Philosophy and Social Studies of Biology (ISHPSSB), University of Guelph.

Orcutt B.C., George D.G., Fredrickson J.A. and Dayhoff M.O. (1982) 'Nucleic acid sequence database computer system' in *Nucleic Acids Research*, 10(1): 157–74.

Orsenigo, L. (1989) *The Emergence of Biotechnology* (London: Pinter Publishers).

Parry B. (2004) *Trading the Genome: Investigating the Commodification of Bio-Information* (New York: Columbia University Press).

Partridge S.M. and Blombäck B. (1979) 'Pehr Victor Edman: 14 April 1916 – 19 March 1977' in *Biographical Memoirs of Fellows of the Royal Society'*, 25: 241–65.

Paul D. (1984) 'Eugenics and the left' in *Journal of the History of Ideas*, 45(4): 567–90.

Pauly P. (2000) *Biologists and the Promise of American Life: From Meriwether Lewis to Alfred Kinsey* (New Haven: Princeton University Press).

Peacock A.C. and Dingman, C.W. (1968) 'Molecular weight estimation and separation of ribonucleic acids by electrophoresis in agarose-acrylamide composite gels' in *Biochemistry*, 7(2): 668–74.

Pederson K. (1983) 'The Svedberg and Arne Tiselius. The early development of protein chemistry at Uppsala' in G. Semenza (ed.) *Selected Topics in the History of Biochemistry: Personal Recollections*, Comprehensive Biochemistry, vol. 35 (New York: Elsevier).

Penders B., Horstman K. and Vos R. (2008) 'Walking the line between lab and computation: the "moist" zone' in *BioScience*, 58(8): 747–55.

Pickstone J. (2000) *Ways of Knowing: A New History of Science, Technology and Medicine* (Chicago: University of Chicago Press).

Pickstone J. (2007) 'Working knowledges before and after circa 1800: practices and disciplines in he history of science, technology, and medicine' in *Isis*, 98: 489–516.

Pickstone (2011a) 'A brief introduction to ways of knowing and ways of working' in *History of Science*, XLIX: 235–45.

Pickstone (2011b) 'Afterwords' in *History of Science*, XLIX: 349–74.

Pierrel J. (2009) *La pratique du séquençage ARN à Cambridge, Strasbourg et Gand, 1960–1980* [The Practice of ARN Sequencing in Cambridge, Strasbourg and Ghent, 1960–1980]. PhD dissertation, Institute for Interdisciplinary Research on Science and Technology (IRIST), University Louis Pasteur, Strasbourg.

Pierrel J. (2012) 'An RNA phage lab: MS2 in Walter Fiers' Laboratory of Molecular Biology in Ghent, from genetic code to gene and genome, 1963–1976' in *Journal of the History of Biology*, 45: 109–38.

Pirie W. (1939) 'Amino-acid analysis and protein structure' in *Annual Report of the Chemical Society*, 36: 351–3.

Podolsky S. and Tauber A. (2000) *The Generation of Diversity: Clonal Selection Theory and the Rise of Molecular Immunology* (Cambridge: Harvard University Press).

Powell A., O'Malley M., Müller-Wille S., Calvert J. and Dupré J. (2007) 'Disciplinary baptisms: a comparison of the naming stories of genetics, molecular biology, genomics and systems biology' in *History and Philosophy of the Life Sciences*, 29: 5–32.

Prober J., Trainor G.L., Dam R.J., Hobbs F.W., Robertson C.W., Zagursky R.J., Cocuzza A.J., Jensen M.A. and Baumeister K. (1987) 'A system for rapid DNA sequencing with fluorescent chain-terminating dideoxynucleotides' in *Science*, 238: 336–41.

Quastler H. (ed., 1953) *Essays on the Use of Information Theory in Biology* (Urbana: University of Illinois Press).

Quirke V. (2007) *Collaboration in the Pharmaceutical Industry: Changing relationships in Britain and France,1935–1965* (London and New York: Routledge).

Rabinow P. (1996) *Making PCR: A Story of Biotechnology* (Chicago: University of Chicago Press).

Rabinow P. (1999) *French DNA: Trouble in Purgatory* (Chicago: University of Chicago Press).

Rabinow P. and Dan-Cohen T. (2005) *A Machine to Make a Future: Biotech Chronicles* (New Haven: Princeton University Press).

Ramillon V. (2007) *Le deux génomiques. Mobiliser, organiser, produire: du séquençage à la mesure de l'expression des gènes* [The Two Genomics. Mobilising, Organising and Producing: From Sequencing to the Measurement of Gene Expression]. PhD dissertation, École des Hautes Études en Sciences Sociales, Paris.

Ramillon V. (2008) 'The network and the factory: organizing control in decentralized sequencing projects', paper presented at the joint meeting of the Society for Social Studies of Science (4S) and European Association for the Study of Science and Technology (EASST), University of Rotterdam.

Rasmussen N. (1999) *Picture Control: The Electron Microscope and the Transformation of Biology in America, 1940–1960* (Stanford: Stanford University Press).

Reardon J. (2005) *Race to Finish: Identity and Governance in an Age of Genomics* (Princeton: Princeton University Press).

Reingold N. (1981) *Science in America, a Documentary History* (Chicago: University of Chicago Press).

Reingold N. (1991) *Science, American Style* (New Brunswick: Rutgers).

Rheinberger H.J. (1993) 'Experiment and orientation: early systems of in vitro protein synthesis' in *Journal of the History of Biology*, 26: 443–71.

Rheinberger H.J. (1995) 'Beyond nature and culture: a note on medicine in the age of molecular biology' in *Science in Context*, 8: 249–63.

Rheinberger H.J. (1997) *Towards a History of Epistemic Things: Synthesizing Proteins in the Test Tube* (Stanford: Stanford University Press).

Rheinberger H.J. (2006) 'The Notions of regulation, information, and language in the writings of François Jacob' in *Biological Theory*, 1(3): 261–7.

Rheinberger H.J. and Gaudillière J.P (eds, 2004a) *Classical Genetic Research and Its Legacy: The Mapping Cultures of Twentieth Century Genetics* (London and New York: Routledge).

Rheinberger H.J. and Gaudillière J.P (eds, 2004b) *From Molecular Genetics to Genomics: The Mapping Cultures of Twentieth Century Genetics* (London and New York: Routledge).

Roberts R. (1993) 'An amazing distortion in DNA induced by a methyltransferase', Nobel Lecture, available at http://www.nobelprize.org/nobel_prizes/medicine/laureates/1993/roberts.html, last accessed January 2012 .

Roberts R. (2000) 'The early days of bioinformatics publishing' in *Bioinformatics*, 16(1): 2–4.

Robertson H.D., Barrell B., Weith H. and Donelson J. (1973) 'Isolation and sequence analysis of a ribosome-protected fragment from bacteriophage ØX 174 DNA' in *Nature New Biology*, 241: 38–40.

Rosenberg D. (ed., 2003) *Early Modern Information Overload*, special issue of the *Journal of the History of Ideas*, 64(1).

Ruhleder K. (1995) 'Reconstructing artifacts, reconstructing work: from textual edition to on-line databank' in *Science, Technology and Human Values*, 20(1): 39–64.

Ryle A.P., Sanger F., Smith L.F. and Kitai R. (1955) 'The disulphide bonds of insulin' in *Biochemical Journal*, 60(4): 541–56.

Saiki R., Scharf S., Faloona F., Mullis K., Horn G., Erlich H. and Arnheim N. (1985) 'Enzymatic amplification of beta-globin genomic sequences and restriction site analysis for diagnosis of sickle-cell anemia' in *Science*, 230: 1350–4.

Sammet J.E. (1969) *Programming Languages: History and Fundamentals* (Englewood Cliffs: New Jersey: Prentice Hall).

Sanderson M. (1972) *The Universities and British Industry, 1850–1970* (London and New York: Routledge).

Sanger F. (1945) 'The free amino groups of insulin' in *Biochemical Journal*, 39: 507–15.

Sanger F. (1949a) 'The terminal peptides of insulin' in *Biochemical Journal*, 45(5): 563–74.

Sanger F. (1949b) 'Some chemical investigations on the structure of insulin' in *Cold Spring Harbor Symposia on Quantitative Biology: Amino Acids and Proteins*, 14: 153–60.

Sanger F. (1959) 'Chemistry of insulin' in *Science*, 129: 1340–4.

Sanger F. (1963) 'Amino acid sequences in the active centres of certain enzymes' in *Proceedings of the Chemical Society*, 5: 76–83.

Sanger F. (1975) 'The Croonian Lecture' in *Proceedings of the Royal Society of London*, 191, B series: 317–33.

Sanger F. (1980) 'Determination of nucleotide sequences in DNA', Nobel Lecture, available at www.nobel.se/chemistry/laureates/1980/sanger-lecture.html, last accessed January 2012.

Sanger F. (1987) 'Reading the messages in the genes' in *New Scientist*, 21st May edition: 60–2.

Sanger F. (1988) 'Sequences, sequences and sequences' in *Annual Review of Biochemistry*, 57: 1–29.

Sanger F. and Tuppy H. (1951a) 'The amino-acid sequence in the phenylalanyl chain of insulin: I. The identification of lower peptides from partial hydrolysates' in *Biochemical Journal*, 49(4): 463–81.

Sanger F. and Tuppy H. (1951b) 'The amino-acid sequence in the phenylalanyl chain of insulin: II. The investigation of peptides from enzymic hydrolysates' in *Biochemical Journal*, 49(4): 481–90.

Sanger F. and Thompson E.O. (1953a) 'The amino-acid sequence in the glycyl chain of insulin: 2. The identification of lower peptides from partial hydrolysates' in *Biochemical Journal*, 53(3): 353–66.

Sanger F. and Thompson E.O. (1953b) 'The amino-acid sequence in the glycyl chain of insulin: II. The investigation of peptides from enzymic hydrolysates' in *Biochemical Journal*, 53(3): 369–74.

Sanger F., Brown H. and Kitai R. (1955) 'The structure of pig and sheep insulins' in *Biochemical Journal*, 60: 556–64.

Sanger F., Hartley B. and Naughton M. (1959) 'The amino acid sequence around the reactive serine of elestase' in *Biochemical and Biophysical Acta*, 34: 243–4.

Sanger F. and Milstein C. (1961) 'An amino acid sequence in the active centre of phosphoglucomutase' in *Biochemical Journal*, 79: 456–69.

Sanger F., Brownlee G.G. and Barrell B. (1965) 'A two-dimensional fractionation procedure for radioactive nucleotides' in *Journal of Molecular Biology*, 13(2): 373–98.

Sanger F., Donelson J.E., Coulson A., Kössel H. and Fischer D. (1973) 'Use of DNA polymerase I primed by a synthetic oligonucleotide to determine a nucleotide sequence in phage f1 DNA' in *Proceedings of the National Academy of Sciences*, 70(4): 1209–13.

Sanger F. and Coulson A. (1975) 'A rapid method for determining sequences in DNA by primed synthesis with DNA polymerase' in *Journal of Molecular Biology*, 94: 441–8.

Sanger F., Nicklen S. and Coulson A. (1977) 'DNA sequencing with chain terminating inhibitors' in *Proceedings of the National Academy of Sciences*, 74: 5463–7.

Sanger F., Air G.M., Barrell B., Brown N., Coulson A., Fiddes J., Hutchison III C., Slocombe P. and Smith M. (1977) 'Nucleotide sequence of bacteriophage ØX174 DNA' in *Nature*, 265: 687–95.

Sanger F., Coulson A., Friedmann T., Air G., Barrell B., Brown L. Fiddes J., Hutchison III C., Sloconbe P. and Smith M. (1978) 'The nucleotide sequence of bacteriophage ØX174' in *Journal of Molecular Biology*, 125: 225–46.

Sanger F., Coulson A., Barrell B., Smith A. and Roe B. (1980) 'Cloning in single-stranded bacteriophage as an aid to rapid DNA sequencing' in *Journal of Molecular Biology*, 143: 161–78.

Sanger F., Coulson, A., Hong G., Hill D. and Petersen G. (1982) 'Nucleotide sequence of bacteriophage λ DNA' in *Journal of Molecular Biology*, 162 (4): 729–73.

Sanger F. and Dowding M. (eds, 1996) *Selected Papers of Frederick Sanger (with Commentaries)* (London: World Scientific).

Santesmases M.J. (2002) 'Enzymology at the core: primers and templates in Severo Ochoa's transition from biochemistry to molecular biology' in *History and Philosophy of the Life Sciences*, 24: 193–218.

Sarkar S. (1996a) 'Decoding "coding": information and DNA' in *BioScience*, 46: 857–63. Reprinted in S. Sarkar (2005) *Molecular Models of Life: Philosophical Papers on Molecular Biology* (Cambridge: MIT Press): 183–204.

Sarkar S. (1996b) 'Biological information: a sceptical look at some central dogmas in molecular biology' in S. Sarkar (ed.) *The Philosophy and History of Molecular Biology: New Perspectives* (Dordrecht: Kluwer): 187–233. Reprinted in S. Sarkar (2005) *Molecular Models of Life: Philosophical Papers on Molecular Biology* (Cambridge: MIT Press): 205–60

Sarkar S. (1998) *Genetics and Reductionism* (Cambridge: Cambridge University Press).

Schaffer S. (1992) 'Late Victorian Metrology and its instrumentation: a manufactory for Ohms', in R. Bud and S. Cozzens (eds), *Invisible Connections: Instruments, Institutions and Science* (Bellingham: SPIE Optical Engineering Press): 23–56.

Schreier P.H. and Cortese R. (1979) 'A fast and simple method for sequencing DNA cloned in the single-stranded bacteriophage M13' in *Journal of Molecular Biology*, 129(1): 169–72.

Scott M., Weiner A.J., Hazelrigg T.I., Polisky B.A., Pirrotta V., Scalenghe F. and Kaufman T.C. (1983) 'The molecular organization of the *Antennapedia* locus of *Drosophila*' in *Cell*, 35: 763–76.

Secord J. (2004) 'Knowledge in transit' in *Isis*, 95(4): 654–72.

Secord J.A. and Jardine N. (eds, 1996) *Cultures of Natural History* (Cambridge: Cambridge University Press).

Segal J. (2003) *Le Zéro et le Un: Histoire de la Notion Scientifique d'Information au 20e Siècle* [*The Zero and the One: A History of the Scientific Notion of Information in the Twentieth Century*] (Paris: Syllepse).

Shannon C. (1948) 'The mathematical theory of communication' in *Bell System Technical Journal*, 28(4): 656–715.

Singer M. (1980) 'Recombinant DNA revisited' in *Science*, 209: 1318.

Sinsheimer R. (1989) 'The Santa Cruz Workshop – May 1985' in *Genomics* 5(4): 954–56.

Sloan P.R. (ed., 2000) *Controlling Our Destinies: Historical, Philosophical, Ethical and Theological Perspectives on the Human Genome Project* (Indiana: University of Notre Dame).

Smith E.L. (1977) 'Amino acid sequences of proteins: the beginnings' in *Annals of the New York Academy of Sciences*, 325: 107–20, special issue on *The Origins of Modern Biochemistry: A Retrospect on Proteins*.

Smith L., Rubenstein J., Parce J.W. and McConnell H. (1980) 'Lateral diffusion of M-13 coat protein in mixtures of phosphatidylcholine and cholesterol' in *Biochemistry*, 19: 5907–11.

Smith L., McConnell H., Baron A.S. and Parce J.W. (1981) 'Pattern photobleaching of fluorescent lipid vesicles using polarised laser light' in *Biophysics Journal*, 33: 139–46.

Smith L., Fung S., Hunkapiller M., Hunkapiller T. and Hood L. (1985) 'The synthesis of oligonucleotides containing an aliphatic amino group at the

5' terminus: synthesis of fluorescent DNA primers for use in DNA sequence analysis' in *Nucleic Acids Research*, 13(7): 2399–412.

Smith L. and Fung S. (1986) 'Deoxyribonucleoside phosphoramidites in which an aliphatic amino group is attached to the sugar ring and their use for the preparation of oligonucleotides containing aliphatic amino groups'. US Patent number 4,849,513, filed 24 June 1986 and issued 18 July 1989.

Smith L., Sanders R.J., Kaiser P., Hughes C., Dodd C.R., Connell C., Heiner S., Kent S. and Hood L. (1986) 'Fluorescence detection in automated DNA sequence analysis' in *Nature,* 321:674–9.

Smith L. and Hood L. (1987) 'Mapping and sequencing the human genome: how to proceed' in *Bio/Technology,* 5: 933–9.

Smith P.R. (1974) 'Reservations on Project K' in *Perspectives in Biology and Medicine,* 18: 21–3.

Smith T. (1990) 'The history of the genetic sequence databases' in *Genomics,* 6: 701–7.

Sommer M. (2008) 'History in the gene: negotiations between molecular and organismal anthropology' in *Journal of the History of Biology,* 41: 473–528.

Spackman D., Stein W. and Moore S. (1958) 'Automatic recording apparatus for use in the chromatography of amino acids' in *Analytical Chemistry,* 30(7): 1190–207.

Staden R. (1977) 'Sequence data handling by computer' in *Nucleic Acids Research,* 4(11): 4037–52.

Staden, R. (1978) 'Further procedures for sequence analysis by computer' in *Nucleic Acids Research,* 5(3): 1013–5.

Staden R. (1979) 'A strategy of DNA sequencing employing computer programs' in *Nucleic Acids Research,* 6(7): 2601–10.

Staden R. (1980) 'A new computer method for the storage and manipulation of DNA gel reading data' in *Nucleic Acids Research,* 8(16): 3673–94.

Staden R. (1982) 'Automation of the computer handling of gel reading data produced by the shotgun method of DNA sequencing' in *Nucleic Acids Research,* 10(15): 4731–51.

Staden R. (1984a) *Computer Methods to Aid in the Determination and Analysis of Nucleic Acid Sequences.* PhD dissertation, MRC Laboratory of Molecular Biology, Cambridge, UK.

Staden R. (1984b) 'A computer program to enter gel reading data into a computer' in *Nucleic Acids Research,* 12(1): 499–503.

Staden R. and McLachlan A.D. (1982) 'Codon preference and its use in identifying protein coding regions in long DNA sequences' in *Nucleic Acids Research,* 10(1): 141–56.

Stadler M. (2010) *Assembling Life. Models, the Cell, and the Reformations of Biological Science, 1920 – 1960*. PhD dissertation, Centre for the History of Science, Technology and Medicine, Imperial College, London.

Star S.L. (2010) 'This is not a boundary object: reflections on the origin of a concept' in *Science, Technology and Human Values,* 35(5): 601–17.

Star S.L. and Griesemer J. (1989) 'Institutional ecology, translations and boundary objects: amateurs and professionals in Berkeley's Museum of Vertebrate Zoology, 1907–39' in *Social Studies of Science,* 19(3): 387–420. Reprinted in M. Biogioli (ed.) *Science Studies Reader* (London and New York: Routledge).

Strasser B. (2002) 'Who cares about the double helix?' in *Nature*, 442: 803–4.

Strasser B. (2003) 'The transformation of the biological sciences in post-war Europe' in *EMBO Reports*, 4(6): 540–3.

Strasser B. (2006a) 'Collecting and experimenting: the moral economies of biological research, 1960s–1980s', in S. de Chadarevian and H.J. Rheinberger (eds) *History and Epistemology of Molecular Biology and Beyond: Problems and Perspectives* (Berlin: Max Planck Institute for the History of Science, Preprint number 310): 105–23.

Strasser B. (2006b) 'A world in one dimension: Linus Pauling, Francis Crick and the central dogma of molecular biology' in *History and Philosophy of the Life Sciences*, 28: 491–512.

Strasser B. (2010) 'Collecting, comparing, and computing sequences: the making of Margaret O. Dayhoff's 'Atlas of Protein Sequence and Structure', 1954–1965' in *Journal of the History of Biology*, 43: 623–60.

Strasser B. (2011) 'The experimenter's museum: Genbank, natural history and the moral economies of biomedicine' in *Isis*, 102(1): 60–96.

Strasser B. and de Chadarevian S. (2011) 'The comparative and the exemplary: revisiting the early history of molecular biology' in *History of Science*, XLIX: 317–36.

Stretton A. (2002) 'The first sequence: Fred Sanger and insulin' in *Genetics*, 162: 527–32.

Sturtevant A.H. (1929) 'The claret mutant type of *Drosophila simulans*: a study of chromosome elimination and of cell lineage' in *Zeitschrift für Wissenschaftliche Zoologie*, 135: 323–56.

Suárez-Díaz E. (2007) 'The rhetoric of informational molecules: authority and promises in the early study of molecular evolution' in *Science in Context*, 20(4): 649–77.

Suárez-Díaz E. (ed., 2007) *Science and representation: a historical and philosophical approach*, special issue of *History and Philosophy of the Life Sciences*, 29(2).

Suárez-Díaz E. (2009) 'Molecular evolution: concepts and the origin of disciplines' in *Studies in History and Philosophy of Biological and Biomedical Sciences*, 40: 43–53.

Suárez-Díaz E. (2010) 'Making room for new faces: evolution, genomics and the growth of bioinformatics' in *History and Philosophy of the Life Sciences*, 32: 65–90.

Suárez-Díaz E. and Anaya-Muñoz V. (2008) 'History, objectivity and the construction of molecular phylogenies' in *Studies in History and Philosophy of Biological and Biomedical Sciences*, 39: 451–68.

Sulston J. (1976) 'Post-embryonic development in the ventral cord of *Caenorhabditis elegans*' in *Philosophical Transactions of the Royal Society of London*, B series, 275: 287–97.

Sulston J. and Horvitz, R. (1977) 'Post-embryonic cell lineages of the nematode *Caenorhabditis elegans*' in *Developmental Biology*, 56: 110–56.

Sulston J., Schierenberg E., White J. and Thompson J. (1983) 'The embryonic cell lineage of the nematode *Caenorhabditis elegans*' in *Developmental Biology*, 100: 64–119.

Sulston J., Mallett F., Staden R., Durbin R., Horsnell T. and Coulson A. (1988) 'Software for genome mapping by fingerprinting techniques' in *Computer Applications in Biosciences*, 4(1): 125–32.

Sulston J., Mallett F., Durbin R. and Horsnell T. (1989) 'Image analysis of restriction enzyme fingerprint autoradiograms' in *Computer Applications in Biosciences*, 5(2): 101–6.

Sulston J., Du Z., Thomas K., Wilson R., Hillier L., Staden R., Halloran N., Green P., Thierry-Mieg J., Qiu L., Dear S., Coulson A., Craxton M., Durbin R., Berks M., Metzstein M., Hawkins T., Ainscough, R. and Waterston R. (1992). 'The *C. elegans* genome sequencing project: a beginning' in *Nature*, 356: 37–41.

Sulston J. and Ferry G. (2002) *The Common Thread: A Story of Science, Politics, Ethics and the Humane Genome* (London: Bantam).

Sutcliffe J.G. (1995) 'pBR322 and the advent of rapid DNA sequencing' in *Trends in Biochemical Sciences*, 20(2): 87–90.

Synge R. and Williams E. (1990) 'Albert Charles Chibnall' in *Biographical Memoirs of Fellows of the Royal Society*, 35: 57–96.

Swann, J. (1988) *Academic Scientists and the Pharmaceutical Industry* (Baltimore: Johns Hopkins University Press).

Tansey E.M. (2008) 'Keeping the culture alive: the laboratory technician in mid twentieth century British medical research' in *Notes and Records of the Royal Society*, 62: 77–95.

Tauber A. and Sarkar S. (1992) 'The Human Genome Project: has blind reductionism gone so far?' in *Perspectives in Biology and Medicine*, 35(2): 220–35.

Taylor P.J. (2008) 'The under-recognized implications of heterogeneity: opportunities for fresh views on scientific, philosophical and social debates about heritability' in *History and Philosophy of the Life Sciences*, 30(3–4): 431–56.

The Editors (1974) 'Editorial policy' in *Nucleic Acids Research*, 1(1): unnumbered page.

Thurtle P. (1998) 'Electrophoretic apparatus' in R. Bud and D.J. Warner (eds), *Instruments of Science: An Historical Encyclopaedia* (New York and London: Garland Publishing): 216–8.

Timmermann C. (2008) 'Clinical research in postwar Britain: the role of the Medical Research Council' in C. Hannaway (ed.) *Biomedicine in the Twentieth Century: Practices, Policies and Politics* (Amsterdam: IOS Press): 231–54.

Valentin F. and Jensen R.L. (2004) 'Networks and technology systems in science-driven fields: the case of European food biotechnology' in M. McKelvey, A. Rickne and J. Laage-Hellman (2004) *The Economic Dynamics of Modern Biotechnology* (Cheltenham and Northampton: Edward Elgar): 167–206.

Venter C. (2007) *A Life Decoded: My Life: My Genome* (New York: Viking).

Von Bertalanffy L. (1933) *Modern Theories of Development: An Introduction to Theoretical Biology* (London: Oxford University Press).

Wada A. (1984) 'Automatic DNA sequencing' in *Nature*, 307: 193.

Wada A. (1987) 'Automated high-speed DNA sequencing' in *Nature*, 325: 771–2.

Warwick A. (2003) *Masters of Theory: Cambridge and the Rise of Mathematical Physics* (Chicago: University of Chicago Press).

Watson J. (1969) *The Double Helix: A Personal Account of the Discovery of the Structure of DNA* (New York: New American Library).

Watson J. (1992) 'A personal view of the project' in D. Kevles and L. Hood (eds) *The Code of Codes: Scientific and Social Issues in the Human Genome Project* (Cambridge: Harvard University Press): 164–73.

Watson J. (2003) *DNA: The Secret of Life* (London: Heinemann).

Watson J. (2010) *Avoid Boring People: Lessons from a Life in Science* (New York: Vintage).

Watson J. and Crick F. (1953a) 'Molecular structure of nucleic acids.: a structure for deoxyribose nucleic acid' in *Nature*, 171: 737–8.

Watson J. and Crick F. (1953b) 'Genetical implications of the structure of deoxyribonucleic acid' in *Nature*, 171: 964–7.

Weatherall M. and Kamminga H. (1992) *Dynamic Science: Biochemistry in Cambridge, 1898–1949* (Cambridge: Wellcome Unit Publications).

Weatherall M.W. and Kamminga H. (1996) 'The making of a biochemist. II: The construction of Frederick Gowland Hopkins' reputation' in *Medical History*, 40(4): 415–36.

Webster F. (1997) 'Is this the Information Age? Towards a critique of Manuel Castells'. *City*, 8: 71–84. Reprinted in F. Webster and B. Dimitriou (eds) *Manuel Castells* (SAGE Publications), vol. III.

White J. (1974) *Computer Aided Reconstruction of the Nervous System of Caenorhabditis elegans*. PhD dissertation, MRC Laboratory of Molecular Biology, Cambridge, UK.

Wieber F. (2006) 'Interplay between molecular biology and computational chemistry: models and simulations in the construction of a computational protein chemistry (1960–1980) in S. de Chadarevian and H. J. Rheinberger (eds) *History and Epistemology of Molecular Biology and Beyond: Problems and Perspectives* (Berlin: Max Planck Institute for the History of Science): 95–103 [preprint number 310].

Wiener M. (1981) *English Culture and the Decline of the Industrial Spirit, 1850–1980* (Cambridge: Cambridge University Press).

Wilkie T. (1991) *British Science and Politics since 1945* (London: Blackwell).

Wilkins (2005) *The Third Man of the Double Helix: The Autobiography of Maurice Wilkins* (Oxford: Oxford University Press).

Wills C. (1991) *Exons, Introns and Talking Genes: The Science behind the Human Genome, Project* (New York: Basic Books).

Wilson D. (2008) *Reconfiguring Biological Sciences in the Late Twentieth Century: A Study of the University of Manchester* (Manchester: Manchester University Press).

Wilson D. and Lancelot G. (2008) 'Making way for molecular biology: institutionalising and managing reform of biological science in a UK university during the 1980s and 1990s' in *Studies in the History and the Philosophy of Biological and Biomedical Sciences*, 39: 93–108.

Wilson M. (1997) *The Difference Between God and Larry Ellison: Inside Oracle Corporation* (New York: William Morrow).

Wright S. (1994) *Molecular Politics: Developing American and British Regulatory Policy for Genetic Engineering, 1972–1982* (Chicago: University of Chicago Press).

Wright S. (1998) 'Molecular politics in a global economy' in A. Thackray (ed.) *Private Science: Biotechnology and the Rise of the Molecular Sciences* (Philadelphia: University of Pennsylvania Press): 80–104.

Wu R. (1994) 'Development of the primer-extension approach: a key role in DNA sequencing' in *Trends in Biochemical Sciences*, 19: 429–33.

Wu R. and Kaiser A. (1968) 'Structure and base sequence in the cohesive ends of bacteriophage lambda DNA' in *Journal of Molecular Biology*, 35: 523–37.

Wu R. and Taylor E. (1971) 'Nucleotide sequence analysis of DNA. II. Complete nucleotide sequence of the cohesive ends of bacteriophage λ DNA' in *Journal of Molecular Biology*, 57: 491–511.

Yi D. (2008) 'Cancer, viruses and mass migration: Paul Berg's venture into eukaryotic biology and the advent of recombinant DNA research and technology, 1967–1980' in *Journal of the History of Biology*, 41: 589–636.

Yockey H., Platzman R. and Quasler H. (eds, 1958) *Symposium on Information Theory in Biology* (New York: Pergamon Press).

Zafra R. (2010) *Un cuarto propio conectado: (ciber) espacio y (auto) gestión del yo* [*A Connected Room of One's Own: (Cyber) Space and Individual (Self) Management*] (Madrid: Fórcola Ediciones).

Zimmerman A.S. (2008) 'New knowledge from old data' in *Science, Technology and Human Values*, 33(5): 631–52.

Zuckerkandl E. and Pauling L. (1965) 'Molecules as documents of evolutionary history' in *Journal of Theoretical Biology*, 8: 357–66.

Zweiger G. (2001) *Transducing the Genome: Information, Anarchy, and Revolution in the Biomedical Sciences* (New York: McGraw Hill).

General Index

Index of Persons-Institutions

Page numbers in bold refer to main entries and in italics to illustrations